Genetic Resources of Horticultural Crops

Dharmendra Singh
Assistant Professor (Plant Breeding and Genetics)
College of Horticulture and Forestry
Central Agricultural University
Pasighat- 791 102 (Arunachal Pradesh)

S. Manivannan
Assistant Professor (Pomology)
College of Horticulture and Forestry
Central Agricultural University
Pasighat- 791 102 (Arunachal Pradesh)

IBDC

International Book Distributing Co.
(Publishing Division)

Published by
INTERNATIONAL BOOK DISTRIBUTING CO.
(Publishing Division)
Khushnuma Complex Basement
7, Meerabai Marg (Behind Jawahar Bhawan)
Lucknow 226 001 U.P. (INDIA)
Tel. : 91-522-2209542, 2209543, 2209544, 2209545
Fax : 0522-4045308
E-Mail : ibdco@airtelmail.in

First Edition 2009

ISBN 81-8189-217-8

Composed & Designed at :
Panacea Computers
3rd Floor, Agrawal Sabha Bhawan
Subhash Mohal, Sadar Cantt. Lucknow-226 002
Phone : 0522-2483312, 9452295008, 9335927082
E-mail : prasgupt@rediffmail.com

Printed at:
Salasar Imaging Systems
C-7/5, Lawrence Road Industrial Area
Delhi - 110 035
Tel. : 011-27185653, 9810064311

Preface

Plant genetic resources are basic requirement in any crop improvement and their importance has increased in the recent years. Genetic resources of horticultural crops now are offered as subject in graduate and postgraduate levels in all horticultural colleges. In view of this, the book "Genetic resources of horticulture crops" has been written primarily to serve as a text book for the students of B.Sc. and M.Sc. (Horticulture) as well as researchers. There is no specific book for genetic resources of horticultural crops. Students are required to go through several horticulture books to have a knowledge on plant genetic resources. It was therefore, thought that students should be provided a comprehensive book as per their course requirements. This book covers, concepts of genetic resources, centre of origin and diversity, plant introduction, plant quarantine, conservation strategies, exploration, collection and exchange of germplasm, DNA fingerprinting, Intellectural property rights, source of resistance/ tolerance for biotic and abiotic stresses, high yield and quality traits in vegetables, fruits, ornamentals, plantation crops, spices, medicinal and aromatic plants.

The authors are thankful to Dr. D. S. Rathore, Formerly Vice-Chancellor to CSK HPKV, Palampur for his help in providing literature and guidance in every step. Authors are also thankful to publishers and authors whose books and journal was consulted and incorporated in the text. We wish to express our grateful thanks to M/s International Book Distributing Company for publishing this book.

It is hoped that this book will serve not only undergraduate but also postgraduate students of horticulture, of various universities as a source of revision. Any suggestions or criticism about the improvement of the book will always be welcome.

Authors

Acknowledgements

The authors are most grateful to the following publishers/authors for whose books and journal used a source material while preparing this text book:

1. B.S. Dhillon, R.K. Tyagi, S. Saxena and G.J. Randhawa (2005). Plant Genetic Resources: Horticultural Crops. Narosa Publishing House, New Delhi, India.
2. Hari Har Ram (1997). Vegetable Breeding - Principles and Practices. Kalyani Publishers, New Delhi.
3. K.V. Peter, (1998) Genetics and Breeding Vegetables, ICAR, New Delhi, India.
4. M.S. Frageria, P.S. Arya and A.K. Chaudhary (2001). Vegetable Crops, Breeding and Seed production (Vol. 1). Kalyani Publishers, New Delhi.
5. H.P. Singh and K.L. Chadha (1993). Genetic resources of citrus: In: K.L. Chadha and O.P. Pareek (eds.). Advances in Horticulture, Vol. 1. Malhotra Publishing House, New Delhi, India.
6. T.K. Bose, L.P. Yadav, P. Pal, P. Das and V.A. Parthasarathy (2002). Commercial Flowers - Vol 1, 2nd edn; Naya Prokash, Calcutta.
7. B.P. Singh and Umesh Srivastava (2004). Plant Genetic Resources in Indian Perspective, Theory and Practices. Directorate of Information and Publication of Agriculture, Indian Council of Agricultural Research, New Delhi, India.
8. P.L. Saroj, B.B. Vashishtha and D.G. Dhandar (2004). Advances in Horticulture. International Book Distributing Co. Lucknow, India.
9. P.K. Ray (2001). Breeding of tropical and subtropical fruits. Narosa Publishing House, New Delhi, India.
10. David F. Bestch (1999). DNA Fingerprinting in Agricultural Genetics Programs, Inc. Edited by Glenda D. Webber, Iowa State University, Ames, Iowa 50011.
11. Indian Horticulture, Directorate of information and publications of Agriculture (DIPA), ICAR, New Delhi. India.

Authors

Contents

Chapter 1

Concept of Genetic Resources

1.1 Introduction

Plant genetic resources are basic raw material for improvement of any crop plant. A wider genetic diversity is fundamental for the development of new variety with good quality and higher yield (Frankel and Hawkes, 1975). This diversity is contained in traditional varieties and modern cultivars, as well as wild relatives of the crop and other wild plant species, special genetic stocks as parental lines and genetic stock with known attributes that can be used now or in the future.

The term genetic resources were coined by O. Fraenkel. The term germplasm has been used most frequently for the sum total of the hereditary materials present in a species.

Horticultural crops comprise fruit and nut crops, vegetables, food legumes, roots and tubers, medicinal, aromatic and ornamental plants. Fruits and vegetables provide nutritional security, as these are rich in vitamins and minerals, which are essential for balanced diet. Horticultural crops are more remunerative than agricultural crops because the return is more from per unit of inputs of land, labour and time. The other important aspects are more employment generation and foreign exchange earning which are very vital for economy of the country.

What are plant genetic resources?

The sum total of hereditary materials present in crop species is known as genetic resources or gene pool or germplasm or genetic stocks.

1.2 Classification of gene pool

The concept of gene pool was given by Harlan and De Wet in 1971. The gene pool may be classified into the three following groups:

1.2.1 Primary gene pool (GP_1)

It consists of cultivated species and its weedy and wild relatives which can be crossed easily for transferring genes and produce fertile hybrids.

1.2.2 Secondary gene pool (GP_2)

It consists of all the biological species that can be crossed with the crop but where hybrids are usually sterile. Gene transfer is difficult but not impossible.

1.2.3 Tertiary gene pool (GP_3)

A tertiary gene pool made up of those species which can be crossed only by using advanced techniques.

Primary gene pool

- The primary gene pool consists of the cultivated forms and usually some wild species that can be intercrossed easily for gene transferring.

Secondary gene pool

- The secondary gene pool which can be crossed with the GP_1. The hybrids are usually sterile but with some fertility.

Tertiary gene pool

- The tertiary gene pool in which crossing is difficult, and gene transfer may be possible through special technique.

1.3 Components of genetic resources

Various components of genetic resources are given below:

1.3.1 Landraces

These are the primitive cultivars or traditional cultivars which are the product of selection carried out by farmers continously over many generations. They have high level of genetic diversity which provides high degree of resistance to biotic and abiotic stresses. The main limitations of landraces are less uniform and low yielders. Landraces were first collected by N. I. Vavilov.

1.3.2 Obsolete or old varieties

Obsolete varieties are developed by modern breeding programme and replaced by new varieties.

1.3.3 Recently released or modern varieties or advanced cultivars

The presently cultivated high yielding varieties are known as modern varieties. They have good source of genes for yield and quality. Such cultivars are major part of working collections.

1.3.4 Advanced breeding lines

Advanced lines are placed under testing trials which are not yet ready for release as a variety. This consist nearly homozygous lines.

1.3.5 Wild forms and wild relatives

These are the important source of resistance to biotic and abiotic stresses. Wild forms are easy to cross with concerned crop species, whereas, wild relatives are much more difficult to hybridize with crops than the wild forms. A large number of wild and weedy species related to crops are available but these remained unexploited to a large extent. Wild species are important resources for desirable genes and rootstocks and some of them can be popularized for direct use as horticultural crops.

1.4 Role of genetic resources in horticultural crops

1.4.1 To improve quality

A number of genetic resources are used in breeding to improve the quality factors in horticultural crops. In general, donors are used in parental combination to obtain high sugar and appealing peel colour in fruits, culinary quality in vegetables and attractive flower colour in ornamentals.

1.4.2 To incorporate tolerance/resistance against biotic and abiotic stresses

The productivity of many commercial cultivars decreases due to biotic and abiotic stresses which can be controlled by incorporating tolerant/ resistant genes present in the genetic resources. In tomato, several wild species namely, *Lycopersicon hirsutum* and *Lycopersicon pimpinellifolium* for fungal resistance, *Lycopersicon chilense* and *Lycopersicon peruvianum* for virus resistance, and *Lycopersicon cheesmanii* for tolerance to adverse environments, have contributed genes for biotic and abiotic stresses.

Chapter 2

Centres of Origin and Diversity of Horticultural Crops

2.1 Introduction

A Russian Plant explorer and geneticist, N.I. Vavilov was one of the first scientists to recognize the importance of plant genetic diversity. Almost one third of the world's species originated in South East Asia and most of the main fruit and vegetable crops come from the East and West Asiatic Centre and Mediterranean centre, while roots and tubers and tropical fruit trees are concentrated in the Central American Centre.

Genetic diversity is the foundation of crop improvement. In order to put agriculture industry of Russia on sound footing, N.I.Vavilov and his associates explored the whole world to collect the genetic diversity of cultivated species. Based on his studies, he recognized and proposed the centers of diversity as centers of origin of cultivated plants.

Vavilov proposed eight main centres of diversity and three subsidiary centres of diversity. He proposed two types of diversity viz., primary and secondary. Centers of diversity were termed as centers of origin by N.I. Vavilov. Later Zeven and Zhukovsky (1975), a close associate of Vavilov, proposed 12 mega gene centres of diversity.

The centers of origin and centers of diversity are not synonyms both have different scientific meanings. Centres of origin mean where crop plants have naturally originated, whereas centres of diversity mean where wide genetic diversity is found. Different species of the same crop may have diversity at different places. For example, different species of yams have diversity in West Africa, South East Asia and Tropical America. Likewise, the same species of the crop may have independent diversity at various places. Various crops have more than one centres of origin and it is very difficult to find out, which one of them is the real centres of origin.

2.2 Centres of diversity

Refers to areas where plants show maximum diversity of forms for that species. Centres of diversity are two types, primary and secondary.

2.2.1 Primary centres of diversity

It refers to geographical area where crop plants are supposed to have originated. In other words, the original places of crop plants which are generally uncultivated areas. Such diversity has maximum number of dominant genes and has normally wild traits.

2.2.2 Secondary centres of diversity

It refers to area where crop plants show considerable diversity but they were not originated there; such areas are known as secondary centre of diversity. Such diversity has lesser genetic diversity compare to primary centre of diversity and has mostly desirable characters.

Initially, Vavilov identified eight centres of diversity and later raised their number to eleven (Table 2.1). Presently, twelve centers of diversity are recognized by Zeven and Zhukovsky (1975) which are given along with the important horticultural crops in Table 2.2.

What are centres of diversity?

It refers to area where crop plants show maximum diversity than any other place of the world.

What are primary centres of divesity?

It refers to geographical area where crop plants are supposed to have originated.

What are secondary centres of diversity?

It refers to area where crop plants show considerable diversity.

Table 2.1. The centres of diversity as proposed by N.I. Vavilov (1926, 1935)

S,No.	Name of centres		Geographical area covered
1.	**The Chineses centre**		Centre and Western China.
	Primary centre	**Secondary centre**	
	Peach, citrus, apricot, walnut, radish, egg plant, cucumber, China tea.		
2.	**The Hindustani centre** Brinjal, cucmber, pepper.		Burma, Assam, Malaya, Java, Sumitra, Philipines but excludes North-West India.
	2a. The Indo-Malayan subcentre Banana, coconut, ginger, nutmeg.		
3.	**The Central Asiatic centre** Pear, apple, walnut, almond, garden pea, garlic, carrot, onion.		Afganistan, Tadjiskistan, Uzbekistan, North West Frontier Provinces and Kashmir.
4.	**The Asia minor centre** Grape, pomegranate, fig, carrot, radish, lettuce, cherry.	Apricot	Interior of Asia minor, Transcausia, Iran and highlands of Turmenistan.
5.	**The Mediterranean centre** Cabbage, broad bean, lettuce,lavender, peppermint.		European countries and Mediterranean region.
6.	**Abyssinian centre** Pea, okra and coffee.	Broad bean (*Vicia faba*)	Ethiopia and Hill country
7.	**The Central American centre** Common bean, pepper, squash, pumpkin, corn, sweet potato, papaya, cashew.		South Mexico and Central America.
8.	**The South American centre** Papaya, guava, sweet potato, potato, lime bean, tomato, pumpkin, marigold. **8a. The Chile centre** Potato, strawberry. **8b. The Brazlian Paraguayan centre** Peanut, passion fruit, rubber tree, pineapple, passion fruit.		Peru, Bolivia, Equador, Colombia, parts of Chile, Brazil and Paraguavy.

Table 2.2. World centers of diversity of important crops (Zeven and Zhukovsky, 1975)

1.	**Chinese centre**	
	Primary centre	**Secondary centre**
	Apricot, loquat, litchi, mulberry, mandarin, peach, kiwi, French bean, cabbage, brinjal, radish, snakegourd, musk melon, tea, chrysanthemum, japanese mint.	Water melon, amaranth.
2.	**Indo-Chinese**	
	Banana, sour orange, sweet lime, mandarin, sweet orange, cassava, snakegourd, Indian spinach, spongegourd, ridgėgourd, bittergourd, cucumber, coffee, cardamom, citronella grass.	Mango, sweet potato.
3.	**Australian centre:** Macadamia nut tree, acacia, eucalyptus.	
4.	**Hindustani centre**	
	Mango, jackfruit, banana, ber, jamun, Indian rubber tree, coconut, egg plant, waxgourd, cucumber, ridgegourd, spongegourd, bittergourd, okra, egg plant, black pepper, turmeric, cardamom, ginger.	Pumkin, watermelon, bottle gourd, sweet pepper, Tea, coconut, Indian rubber.
5.	**Central Asian Centre**	
	Pear, apple, walnut, almond, apricot, grape, peach, apple, plum, walnut, pistachio, pea, carrot, radish, onion, garlic, spinach, muskmelon, broad bean, jasmine.	Egg plant, watermelon, cauliflower, apricot, peach.
6.	**Near Eastern centre**	
	Almond, fig, apricot, grape, pomegranate, peach, citrus, almond, grape, pea, melon, cabbage, cucumber, pea, lettuce, onion, fenugreek, coffee, rose.	Garlic
7.	**Mediterranean centre**	
	Cabbage, lettuce, radish, black cumin coriander, cumin, marigold, gladiolus.	Onion, garlic.

8.	**African centre**	
	Date palm, cashew, bottlegourd, cowpea, eggplant, watermelon, cowpea, okra, coffee, gladiolus.	Cassava, onion, lima bean, amaranth.
9.	**European Siberian centre**	
	Apricot, apple, almond, walnut, pear, lettuce, brussels sprout, asparagus potato, bell pepper.	
10.	**South American centre**	
	Papaya, cashew, pineapple, custard apple, passion fruit, sweet potato, potato, tomato, pumpkin, common bean, capsicum, tomato, lettuce, bougainvillea.	Onion, common bean.
11.	**Central American and Mexico centre**	
	Guava, grape, papaya, pine apple, avocado common bean, pepper, squash, potato, cassava, tomato, pumpkins, common bean, amaranth, zinnia.	
12.	**North American centre**	
	Grape, tomato, eggplant, watermelon, pumpkin and squash, onion lettuce, common bean, okra, pepper, mint.	

2.3 Micro-centres

Micro-centres are small areas within the large centres of diversity exhibiting much greater diversity than the centres as a whole. The evolutionary process proceeds at a much faster pace in the micro-centres which makes them more important for diversity collection as well as for studying the plant evolution.

What are micro-centres?

Micro-centres are small areas within the large centres of diversity exhibiting much greater diversity than the centres as a whole.

2.4 Law of homologous series in variation

N.I. Vavilov collected the global genetic diversity of the cultivated plant species and studied the inheritable variation systematically. He found parallel variation of characters among species and genera meaning thereby concluding that the characters of closely related species and genera resemble each other or the same characters occur in them. In other words, characters occur in one species also found in other related species. For example, diploid and tetraploid *Brassicas* show a series of similar characters.

What is law of homologous series in variation?

Close resemblance is found for pattern of variation between the related species so that if any particular character found in one species, the same is also present in the other related species.

Chapter 3

Plant Introduction

3.1 Introduction

Plant indroduction is the process of introducing a plant genotype of a species, variety or wild relative to growing locality to a new locality, it is called plant introduction. The introduced plant genotype may be one or more and each genotype will be considered as a separate plant introduction. Plant introduction may take place between continents, countries or regions of the same country. Introduction of plants from abroad is called foreign introduction or exotic collection. Introduction from one state to another is called indigenous introduction.

Plant Introduction: Introduction of a plant genotype to a new environment is termed as plant introduction.

3.2 Types of plant introduction

Introductions are of two types, primary and secondary.

3.2.1 Primary introductions

When an introduced genotype is released for commercial cultivation in its original form in a new environment, it is termed as primary introduction. Large number of horticultural introductions are primary introductions because these are being grown in the country as such without any further improvement. Introductions like Sioux (USA) and La Bonita of tomato (USA), Bragg (USA) and Lee (USA) of soybean, Contender (USA) and Kentucky Wonder (USA) of frenchbean, Golden Acre (Denmark) and Drum head (Denmark) of cabbage are some examples of primary introduction.

Primary introductions: When an introduced genotype is released for commercial cultivation in its original form in a new region, it is termed as primary introduction.

3.2.2 Secondary introductions

When the introduced genotypes are improved by hybridizing with the local cultivars and released for commercial growing. Such an introduction constitute secondary introduction. Examples of secondary are Aseem (Pusa Phalguni x EC21622) and Rituraj (Pusa Dofasali x EC26410 Mexico) of cowpea, G_2 (Pusa Sawani x Ghana Red) and Parbhani Kranti (Pusa Sawani x Ghana collection) of okra, Pusa Snowball-1 (EC12012 x EC12013) of cauliflower, Pusa Alankar (EC27060 x Sel No IPL 8) of squash, Pusa Ruby (Sioux x Improved Meeruti), Pusa Early Dwarf (Red Cloud x Improved Meeruti), Pusa Kesari (EC65055 x Punjab Tropic) and Punjab Chhuhara (EC55055 x Punjab Tropoic) of tomato, Pusa Kranti (Pusa Purple long x Hyderpur x Wyland Giant) and Pusa Anmol (Pusa Purple long x Hyderpur) of brinjal, Pusa Kanchan (Local red round x Golden Ball) of turnip, Pusa Yamdagaini (EC9981 x Nantes) and Pusa Kesar (Local red x Nantes Half long) of carrot, Pusa Sarbati (Kutana x Resistant No.6) of musk melon' Pusa Alankar (EC27050x Sel NoI PL 8) of squash and Arka Manik (Local selection x Crimson sweet) of water melon.

Secondary introductions: Introduced genotypes are improved by hybridizing with local cultivars and released for commercial cultivation.

3.3 Nodal agency of plant introduction in India

Realizing the importance of plant introduction, ICAR sanctioned a plant introduction scheme at the Division of Botany, Indian Agricultural Research Institute, New Delhi, in 1946. The scheme was again expanded as the Plant Introduction and Exploration Organisation in 1956 which was upgraded as the Division of Plant Introduction of the Institute in 1961. In order to keep pace with the plant genetic resources activities at global level, the Division of Plant Introduction was reorganized in 1976 as a full fledged institute of the ICAR with the name of National Bureau of Plant Genetic Resources (NBPGR). NBPGR is the sole national agency for introducing plant genetic resources in the country. The NBPGR operates from its headquarters at Pusa Campus; New Delhi with five Regional Stations (Akola, Jodhpur, Shillong, Shimla and Trichur) located in different agroclimatic zones of India.

3.4 Objectives of plant introduction

Plant introduction is mainly done with two objectives. First to use the introduced genotype as a donor in a crop improvement programme. Secondly, for releasing the introduced material for commercial cultivation after adaptation.

3.5 Procedure of plant introduction

Introduction involves procurement, quarantine, cataloguing, evaluation, documentation, multiplication and distribution. Plant introduction work is handled by the Divisions of Germplasm Exchange and Plant Quarantine of the NBPGR.

3.5.1 Procurement

Any institute or person may introduce germplasm in India. The requirement of genetic resources is submitted by a scientist to the NBPGR which tries to meet the requirement from its own sources or other sources in the country. If the plant material is not available in the country and is to be obtained from abroad, the NBPGR issues an import permit on the prescribed proforma. The plant material is obtained free of cost on mutual exchange basis through counter-part agencies of other countries.

The plant part to be introduced may be seed, tuber, bulb, runner, sucker, stolon, bud, rooted cutting, scion stick, plant seedling or *in vitro* meristem culture depending upon the method of propagation of the crop. The plant part used for the propagation of a species is called propagule.

The indent for the introduction is sent to the supplier with the details of the genetic resources indicating the period of delivery and import permit. He is also requested to enclose a phytosanitary certificate with the plant materials.

3.5.2 Quarantine

All plant introductions should be routed through plant quarantine and are assigned import quarantine (IQ) number, source name, type of material, case number, consignee, quarantine history etc. It should be recorded in import quarantine register along with the information on clearance/rejection, interceptions, post-quarantine treatment, etc. These records are maintained in the import register. Details see in chapter 4.

3.5.3 Cataloguing

The material is received personally by the officials of the NBPGR on getting the information of its arrival and is catalogued. Each genetic resources is called an accession and is given an accession number with a prefix of either E C (Exotic collection), or I C (Indigenous collection) or I W (Indigenous wild). The information recorded for identification includes name of the species, cultivar, nature of propagule, address of supplier, and important characteristics. Immediately, it is handed over for quarantine inspection. If it is cleared by the quarantine, it is passed on to the indenter. The material is destroyed if it does not confirm the quarantine rules.

3.5.4 Evaluation

The germplasm utilization requires accurate evaluation, characterization, and documentation of the material. Evaluation refers to assessment of germplasm for morphological, physiological, biochemical, entomolical and plant pathological traits.

3.5.5 Documentation

Documentation is one of the important activities of plant genetic resources. After plant quarantine inspection/clearance, introduced germplasm is documented in national register alongwith available passport data and EC number assigned to each collection. Generally it includes donor name, institute address, source country, taxonomic status like, common name, genus, species, cultivar name, pedigree, local name, collector's name, habitat/location, latitude/longitude, specific characteristics, and form of material-seeds, cuttings or plant, destination/recipient. The role of documentation is given below:

(i) providing a tool for an efficient management

(ii) giving the possibility of choosing the appropriate material to use in breeding and research programmes

(iii) providing information to co-ordinate global activity on genetic resources.

3.5.6 Multiplication and Utilization

Promising genotypes are multiplied and supplied to the users on demand for utilization in the crop improvement programme.

3.6 Purpose of plant introduction

The main aim of plant introduction is to serve basic or parent material for genetic improvement. The following purposes of plant introduction are given below:

3.6.1 To introduced new best varieties

The best varieties like Red Chief of apple, Santa Rosa of plum, Tommy Atkins of mango and Ramnos-kaya of sugar beet are made available to the farmers by plant introduction.

3.6.2 To introduce new crops

New crops life guava, papaya and tomato crops are introduced in the recent past and kiwifruit and Japanese mint introduced in the recent times.

3.6.3 To utilize genetic resources for crop improvement

Introduced genetic resources are used for hybridization with local varieties to develop improved varieties. A large number of commercial varieties of cabbage, cauliflower and tomato have been evolved by using genetic resources. In tomato, Pusa Ruby derived from across between Meeruty and Sioux an introduction from U.S.A.

3.6.4 To obtain donors for resistance/tolerance against stresses

Biotic and abiotic stresses are the main causes of reducing the productivity of the crops. Hence, the breeders lay great emphasis on breeding the varieties resistant/tolerant to biotic and abiotic stresses and require donors against these stresses. Several lines of *Lycopersicon pimpinellifolium* have been found resistant to *Fusarium* wilt, fruit rot and early blight which may be used in breeding programme of tomato.

3.6.5 To bring desirable rootstocks

In fruits, the rootstocks have a very significant role in commercial orcharding as they control the vigour of the plants and provide tolerance against adverse soil conditions. In grape, root stock of "Dog Ridge" is a notable example for salt and drought tolerance.

3.7 Merits of plant introduction

Plant introduction has several merits which are presented below:

3.7.1 Quickest method of crop improvement

A very long period is required to evolve desirable cultivars of a crop whereas desirable cultivars available anywhere in the world may be utilized quickly for the development of horticulture by simply procuring it. Thompson Seedless cultivar is responsible for the progress of grape growing in the entire world.

3.7.2 Most economical method of crop improvement

Plant introduction is very economic method of crop improvement particularly when the introductions are released as varieties either directly or through hybridization.

3.7.3 Increases the genetic diversity

An increase in the range of genetic diversity provides the farmer alternatives to choose the cultivars according to his requirements of local environments and market.

3.8 Demerits of plant introduction

Major demerits of plant introduction are associated with the entrance of new insects, pests, diseases, nematodes and weeds alongwith introduced materials. Earlier there was no plant quarantine regulations in force. Hence, following demerits were more common:

3.8.1 Entrance of noxious weed

In the past, many noxious weeds have entered in the country along with crops such as *Argemone mexicana and Phlaris minor. Lantana camera* introduced as an ornamental plant has become a noxious weed.

3.8.2 Insect pests

Woolly aphis of apple and tuber moth of potato entered in India along with the introduced varieties.

3.8.3 Diseases

Late blight of potato and bunchy top of banana are very serious diseases which came in the country along with the introduced plant material.

3.9 Sources of plant introduction

The seed /planting material can be introduced from centres of diversity, international institutes, International board for plant genetic resources, gene banks, national arboreta, botanical gardens, universities/ private nurseries etc.

3.10 Promising introduction in horticultural crops

The number of introduced varieties is very large. Hence, only few are given in fruits and vegetable.

3.10.1 Fruits

(i) **Apple:** Red Delicious, Golden Delicious.

(ii) **Pear:** Bartlett, Flemish Beauty.

(iii) **Peach:** Elberta, Floradasun.

(iv) **Plum:** Santa Rose, Green Gage.

(v) **Apricot:** New Castle, Royal, Nugget.

(vi) **Citrus:** Kinnow, Valencia Late.

(vii) **Grape:** Thompson Seedless, Perlette, Beauty Seedless.

3.10.2 Vegetables

(i) **Tomato:** Sioux, Marglobe, La Bonita.

(ii) **Pea:** Arkel, Bonneville.

(iii) **Onion:** Early Grano, Ratnar Selection, Brown Selection.

(iv) **Cabbage:** Golden Acre, Drum Head.

(v) **Cauliflower:** Snow Ball-16.

(vi) **Radish:** China Red, Japanese White.

(vii) **Turnip:** Golden Ball.

(viii) **Okra:** Ghana Red, Velvet Green, Perkin's Green.

(ix) **Carrot:** Nantes, Nantes Half Long.

(x) **French bean:** Contender, Kentucky Wonder.

Chapter **4**

Plant Quarantine

4.1 Introduction

Plant quarantine is vital to prevent the entry of new diseases, insects and weeds into a country or eradicate them before they can become widespread and well established.

What is plant quarantine?

It refers to prevention of the entry of new diseases, insects and weeds.

4.2 Principles and concepts of plant quarantine

The aim of quarantine is to check the introduction of exotic pests and diseases and to prevent the spread of those already present in the country from one region to another. The plant material is imported by breeders to utilize in the crop improvement programmes and also by farmers for commercial cultivation. Hence, it is necessary to adopt proper quarantine measures to prevent the entry of pests, diseases and weeds with the imported plant material. All the countries are facing the danger of entry of serious pests and diseases with the current large scale exchange of plant materials for the development of agriculture.

Quarantine means the practice of strict isolation and inspection of the plant material as a safeguard against spreading of pests and diseases. The governments of all countries have enacted laws to enforce it. Plant quarantine may be defined as government endeavour, through legislative measures, to regulate the introduction of plants, planting materials, plant products, soil, living organisms, etc. Prevention of inadvertent introduction of pests, pathogens and weeds harmful to the agriculture of a country/state/region and if introduced, check their establishment and further spread.

4.3 Plant quarantine regulations

Plant quarantine regulations are promulgated to prevent the introduction and spread of harmful pests and pathogens. The plant quarantine is of two categories, i.e., foreign and domestic plant quarantine. The first quarantine law was passed in Rouen (France) in 1890 to prevent the spread of bar berry the alternate host plant of wheat rust. The import and export of seeds, plants, plant products and planting materials in India are regulated by the rules and regulations framed under the Destructive Insects and Pests Act (DIP Act). Such act issues various notifications, imposes the quarantine restrictions on plant materials being imported or moving from one state to another in the country. The seed was not originally included in the DIP Act but because of the changing situations and to meet current requirements, the Govt. of India passed a comprehensive 'Plants, Fruits and Seeds Order, 1984' which came into force in June 1985. The objective of this act is to prevent the entry of new diseases, insects, pests, weeds etc. into the country.

Introduction of genetic resources of horticultural crops is done mostly through vegetative propagules such as rooted cuttings, unrooted cuttings, scionwood, bulbs, rhizomes, corns, and plants which carry more pests and diseases than seeds. The following quarantine regulations are mostly enforced:

4.3.1 No consignment shall be imported into India without an official 'Import Permit' issued by the Plant Protection Adviser, Government of India, Ministry of Agriculture.

4.3.2 No consignment can be imported into India without an official 'Phytosanitary Certificate' issued by the official plant quarantine agency of the exporting country.

4.3.3 All consignment of plants and seeds shall be imported into India through seaport and airport.

4.3.4 Soil, sand and plant debris accompanying seeds/planting materials shall not be permitted to import. But soil can be imported for research purpose if permit issued by the Plant Protection Adviser, Government of India, Ministry of Agriculture.

4.4 Plant quarantine set up

The Directorate of Plant Protection, Quarantine and Storage, Faridabad headed by the Protection Advisor, Govt. Of India framed under the DIP Act. Such Directorate has been entrusted with the responsibility of enforcing the quarantine rules and regulations. For this purpose, plant quarantine stations have been established. Elaborate facilities including post entry quarantine green houses are also being developed at Amritsar, Delhi, Mumbai, Kolkata and Chennai. The National Bureau of Plant Genetic Resources, New Delhi has been authorized to deal with the quarantine requirements of germplasm materials for research purpose. Its Division of Plant Quarantine is performing the functions of plant quarantine effectively.

4.5 Plant quarantine agencies in India

The Government of India has approved three National Organizations/ Institutes to look after official quarantine agencies for exchange of seed/planting material for different group of crops/plants, especially for research purposes:

4.5.1 The National Bureau of Plant Genetic Resources (NBPGR), New Delhi

The institute looks after the quarantine inspection works of agri-horticultural and agri-silviculture plants.

4.5.2 The Forest Research Institute (FRI), Dehradun (Uttaranchal):

Forest Research Institute deals with the quarantine inspection of forest tree propagules.

4.5.3 The Botanical Survey of India (BSI), Kolkata (W.B.)

Botanical Survey of India looks after the quarantine inspection of the plants of the botanical interest.

Chapter 5

Exploration and Collection of Germplasm

5.1 Introduction

World heritage of biodiversity has been diminishing at an alarming rate during the last four decades of the twentieth century due to replacement of landraces with modern cultivars, deforestation, population pressure and urbanization. There is an argent need to collect and conserve the genetic diversity for use in present as well as in future. To collect the biodiversity, explorations are conducted. Explorations are therefore, expeditions made to the regions of diversity in order to collect the samples of variability for conservation and utilization.

5.2 Objectives of exploration

Plant exploration has two main objectives. Firstly, to collect maximum amount of diversity including cultivated and wild forms and secondly, to understand the nature and pace of evolutionary process by studying the variation within and between the species.

5.3 Types of exploration

Explorations are of two types, crop specific (specific exploration) and region specific (broad based exploration).

5.3.1 Crop specific exploration

It is undertaken to collect variability in a particular crop or cultigen or material of specific attributes.

5.3.2 Region specific exploration

It is undertaken to tap maximum diversity in different crops occurring in the region.

5.4 Exploration planning

A meticulous planning is required for making an exploration successful because it involves lot of expenditure and time. The planning should begin many months in advance by making contacts with the local officials, consultation of the published flora of the region, herbarium, knowledge of agro-ecology and social customs and raising of rootstocks for propagation of fruits. Subsequently, it may be supplemented with the arrangements for transport, equipments, communication and camping. The following logistics preparations are required for exploration of horticultural crop:

5.4.1 The exploration team should consist of horticulturist crops curater, botanist and a local guide.

5.4.2 The expedition should coincide with the period of crop harvest so as to evaluate the variability on the spot as well as to collect the seeds in vegetables.

5.4.3 The duration depends on many factors. However, it should be 4 weeks within the country and 2 to 3 months in a foreign country.

5.5 Exploration strategies

The collection needs are determined on the basis of degree of threat to erosion, adequacy of existing collections, need of the breeders and researchers and socio-economic importance of the species.

Efforts should be made to collect the entire gene pool by sampling the whole area of distribution of the species including the periphery where species is under stress. In case of fruits, the exploration has to be done at least in two phases. Firstly, when harvesting of the fruits has just begun which helps in identification of diversity. Secondly, during the propagating period to collect the scion wood. The genetic resources of other horticultural crops can be collected during the harvest period. Explorer should have the following information:

5.5.1 Good knowledge about the diversity of varieties, primitive, land races and local types etc.

5.5.2 Knowledge of wild realtives species and distribution of their variability.

5.5.3 Adequate knowledge of sampling method, sample size, varietal characters.

5.5.4 Knowledge of appropriate propagation method/time and availability of sufficient root stocks.

5.6 Exploration methodology

5.6.1 Coarse v/s fine grid sampling

Initially, a coarse grid random sampling is done from the entire region of diversity whereas, fine grid random sampling is done from the concentrated pockets of variation, *i.e.* gene microcentre.

5.6.2 Random v/s biased sampling

The sampling strategy depends on the crop species self or cross pollinated or vegetatively propagated species. Both are important in horticultural crops. Random sampling is more practiced in the crops raised by seeds in order to collect maximum genetic diversity. But in case of limited availability of the fruits and vegetables, it is not possible to collect large samples randomly. Further, a specific genotype has to be collected in fruits because it is not possible to reconstruct that genotype through the randomly collected variability. Thus, biased sampling has a significant role in horticultural explorations.

5.7 Collection of germplasm

Collection refers to obtaining the various germplasm accessions from various sources. The purpose of germplasm collection is to reduce the loss of genetic diversity and make available for future use. The following points should be considered at the time of germplasm collection :

5.7.1 Germplasm can be collected from centres of diversity and gene bank. In addition, collection should also be made from gene sanctuary and farmers field because farmers especially in tribal areas still grow obsolete cultivars which have considerable genetic diversity.

5.7.2 The priority of collection will be decided by the need of the breeders. But endangered species should be given priority for collection of germplasm.

5.7.3 Random as well as nonrandom sampling should be done for tapping of genetic diversity of a crop species.

5.7.4 It has been recommended that 50-100 plants should be collected from each sample site with 50 seeds from each plant. This sample size is sufficient for capturing maximum genetic diversity.

Why collection of germplasm is necessary?

Collection of germplasm is necessary for tapping of genetic diversity.

5.8 Passport data for collection of germplasm

Passport data includes all basic information of the site from where sample has been collected. It includes ecological and habitat data, altitude and climate etc. Such information is very useful for all phases of genetic resources work. Site identification is perhaps the most significant evidence available to the curator for designating a 'core collection' and is of great help in identifying duplicates (Frankel, 1989). Records on topography or soil characteristics can be valuable for plant breeders for improving adaptation to particular conditions especially of tolerance for edaphic or climatic stresses.

Each accession is assigned an IC (Indigenous Collection) number before it is passed on to the Evaluation Division. In the past, the passport data information was documented manually in the form of Plant Collection Reporter and disseminated to user agencies. Recently, some progress has also been made to computerise such information. The following information is required for collection of germplasm.

5.8.1 Accession data

(i) Accession number
(ii) Donor name
(iii) Donor identification number
(iv) Other numbers associated with accession
(v) Scientific name
(vi) Name/cultivar/pedigree
(vii) Acquisition date
(viii) Accession size.

5.8.2 Collection data

(i) Collection number
(ii) Collecting institute
(iii) Date of collection of original sample
(iv) Site, country, state, district, village and locality
(v) Latitude and longitude of collection site
(vi) Altitude of site
(vii) Collection source: wild, farm land, store, market, institutes etc.
(viii) Local/vernacular name

(ix) Topography
(x) Associated crops
(xi) Cropping density
(xii) Pests
(xiii) Diseases
(xiv) Remarks.

Chapter **6**

Exchange of Germplasm

6.1 Introduction

Import and export of germplasm constitutes germplasm exchange. Germplasm import is the major activity in plant introduction.

What is germplasm exchange?

Import and export of germplasm materials from all available sources is known as germplasm exchange.

6.2 Procedure for exchange of germplasm

The procedure for exchange of germplasm is given below :

6.2.1 Import

Indentors are to be made requests to the NBPGR for import germplasm from abroad. Details of the required material, source/ country as well as address of organisation/scientist may be given in a prescribed application.

The concerned scientist/organisation abroad is advised to take into consideration the following requirements for mailing the material to India.

(i) Only healthy materials have to be forwarded without any seed treatment so as to facilitate proper quarantine examination may be done.

(ii) Packages of seed/planting material should be addressed to the Director, NBPGR, New Delhi, who has to take delivery of the seed/planting materials and conduct quarantine examination. The material is required to be accompanied with an 'import permit' and 'phytosanitary certificate' with additional declaration, if any. Two copies of the phytosanitary certificate are required for one inside the package and the other affixed outside for use by the quarantine authorities.

(iii) The seed/planting materials should not be sent to any scientist by name directly. Such materials should be properly monitored for quarantine requirements for documentation of passpórt data and national accessioning.

(iv) The perishable plant propagules (scion/woods, budwoods, plant rhizomes, suckers, etc.) may preferably be sent by airfreight through any commercial airlines.

(v) The sender should be provided full details of the seed/planting material along with the address of the concerned scientist.

6.2.2 Export

The following guidelines are to be followed for export of germplasm:

(i) Requests received from concerned organisations/agencies abroad have to be forwarded to the NBPGR for seed/planting material with relevant information.

(ii) Seed/planting materials should be desptched through the Bureau, so that inspection of the seeds/ material could be done from quarantine angle and phytosanitary certificate be issued.

(iii) Only the healthy materials should be sent to the Bureau in small quantity along with full details of the material(s) and the name and address of the recipient. Quarantine process normally takes about 7 to 10 days.

Chapter 7

Germplasm Conservation

7.1 Introduction

Plant germplasm conservation or genetic conservation means preservation of plant genetic resources for use in the present as well as in future.

A greater need of genetic conservation has arisen because of the prevailing genetic erosion, genetic vulnerability and acceleration use of plant genetic resources to increase horticultural productivity. Genetic diversity is decreasing rapidly in the regions of the centres of diversity as well in the cultivated areas. The land races or primitive varieties are being replaced everywhere with modern high yielding cultivars. The other causes of loss of genetic diversity are deforestation, land clearance, population pressure, urbanization and shifting land use patterns. Hence, there is an urgent need to conserve the genetic diversity for current and future needs. The conservation and utilization of plant genetic resources is the key to improving agricultural productivity and sustainability thereby contributing to national development, food security and poverty alleviation.

What is germplasm conservation ?

Preservation of genetic diversity of crop plants is termed as germplasm conservation.

Why are plant genetic resources conserved ?

Conservation of genetic resources secures the availability of biodiversity for farmers, breeding and research and for the use of future generation. Adequate genetic diversity is particularly important for breeding and thereby for food security of India.

What is genetic erosion?

Gradual reduction in genetic variation from cultivated species and their wild forms and relatives is termed as genetic erosion.

7.2 Conservation strategies

There are two basic approaches to the conservation of plant genetic resources namely *ex-situ* and *in-situ* conservation :

7.2.1 *Ex-situ* conservation

Ex-situ conservation is the preservation of species and populations outside the native habitat in the form of seeds, embryos, tissues or plants. The objective of *ex-situ* conservation is to maintain the accession without change as regards to its genetic constitution (Frankel and Soulé 1981).

What is *ex–situ* conservation ?

Conservation of germplasm is attempted away from its natural habitat, is known as *ex-situ* conservation.

7.2.1.1 Technologies and methods for *ex-situ* conservation

Approach to *ex-situ* conservation includes seed gene banks, field gene banks and *in vitro* conservation. Among the various *ex-situ* conservation methods, a seed gene bank is the most convenient for long term conservation of plant genetic resources.

7.2.1.1.1 Seed gene banks

Germplasm is stored as seeds of various accessions. Seeds are the most convenient form for the genetic conservation and the importance of seed storage has been recognized ever since humans began to domesticate plants. From the view point of storage, seeds have been grouped into two categories, orthodox and recalcitrant (Robert, 1973). Orthodox seeds can be dried, without damage, to low levels of moisture content. The horticultural crops having orthodox seeds are carrots, beets, onion, papaya, pepper, cucumber, squashes, tomato, beans, eggplant, spinach and all *Brassicas.* On the contrary, recalcitrant seeds are unable to withstand much desiccation and can be stored only in wet medium in order to avoid dehydration injury. They remain viable only for a short time (weeks or months), even if kept in the required moisture condition. The horticultural crops producing recalcitrant seeds include citrus, mango, rubber, chestnut, walnut, avocado, jackfruit, coconut, oil plam, coffee, cocoa, and coconut, litchi, tea, cardamom and plantation crops. However, some of them (citrus, oilpalm, and coffee) can be stored for a longer period with the adjustments of temperature and humidity. Generally seed gene bank collections are classified into three groups.

(i) **Base collections**

Plant materials can be stored for long periods of time (over 20 years) are known as base collections. The genetic resources are not distributed to users from the base collection. Seeds are stored at temperature of -10 to -20° C and seed moisture content is kept at 5 ± 1 per cent.

(ii) **Active collections**

Accessions are stored for short to medium periods of time (10-15 years) as is often the case for breeder's collections, for regeneration, evaluation, research and distribution to end users. Active collections should be kept at 0^0 C temperature with moisture content around 8 per cent. Active collections are the accessions which are immediately available for multiplication, distribution and use.

(iii) **Working collections**

These collections may be stored for 3-5 years and no need to be grown every year. Seeds are stored at temperature of 5-10^0 C and seed moisture content is kept at 8-10 per cent. Plant breeders maintain accessions for selection and crossing purposes as part of breeding programmes.

What are Orthodox seeds?

Orthodox seeds can tolerate drying to low moisture content of 5% or lower. The viability can be maintained by drying the seeds and storing these at low temperature for very long period.

What are recalcitrant seeds?

Recalcitrant (desiccation intolerant) seeds which show very drastic loss in viability if their moisture content is below a critical limit usually between 12-30%.

7.2.1.1.2 Field gene banks

Field genebank is the most important method of genetic resources conservation for a large number of horticultural crops which are propagated vegetative and also their seeds are recalcitrant. The genetic resources of such crops are preserved in field genebanks which are also known as clonal repositories, clonal materials or living collections. Field genebank is a facility where clonal genetic resources are conserved as living collections in field, orchard or plantation. Field gene banks have two objectives. The primary objective is to make available the authentic genetic resources of a conserved crop to the breeders as plant parts and fruit samples. The secondary objective is to conduct research on evaluation, propagation, conservation, cytogenetics and biosystematics. The main function of a field genebank is to maintain the collected genetic resources under proper cultural and phytosanitory conditions.

7.2.1.1.3 *In vitro* conservation

The genetic resources of a large number of horticultural crops are maintained in field genebanks. Field genebanks require a lot of land and are very expensive to maintain and vulnerable to germplasm losses caused by pests, diseases and natural disasters. *In vitro* conservation has been advocated by many scientists as an alternate or complement of field genebank.

In-vitro conservation refers to maintenance of germplasm in a defined and artificially creative environment. Slow growth *in vitro* can be done in several ways such as minimal media by reduction in sucrose or the macro or micro nutrients, thus restricting the availability of nutrients to culture. Besides, incorporation of osmotic agents such as Sorbitol or Mannitol also helps in reduction in growth rate. Large collections of banana, cassava and potato are being conserved by this technique.

7.2.1.1.4 Cryopreservation

Cryopreservation involves preservation of plant material at very low temperature in liquid nitrogen (-196^{0}C). At this temperature all cellular division and metabolic events remain suspended and the material can be stored without changes for long period of time. Cryopreservation of plant material is the only viable option for long term storage of vegetatively propagated species or species with recalcitrant seeds. It can also be used for the species of orthodox seeds. The cryopresevation has been successfully applied to callus, protoplasts, shoot apices, somatic embryos, seeds and excised zygotic embryos.

The following steps are involved for cryopreservation techniques (Engelmann, 1991):

(i) **Choice of material**

As a general rule, young and meristmatic material should be used for cryopreservation technique.

(ii) **Cryoprotective substances**

It is carried out using various cryoprotective substances like mannitol, sucrose, sorbitol, dimethyl sulphoxide, polyethylene glycol etc. For every species, one will have to be determine the choice of cryoprotectants.

(iii) **Freezing**

Different types of freezing process can be categorized as ultra rapid, rapid or droplet freezing.

(iv) **Storage**

Samples are permanentaly kept at or near the temperature of liquid nitrogen (-196^{0}C).

(v) **Thawing**

In the majority of the cases, thawing is found suitable for most of the cryopressure protocols. It is accomplished by removing the specimen from liquid nitrogen and immersing it in warm (34-40^{0}C) water bath for 1-2 minutes.

(vi) **Regrowth**

Regrowth of cryopreserved specimen is the most accurate criterion for assessing viability as compared to other methods such as vital staining or tetrazolium chloride tests.

7.2.1.1.5 New cryopreservation techniques

The folllwing new techniques are available for cryopreservation :

(i) Vitrification

Explants are placed in the presence of a highly concentrated cryoprotective solution then freezing them rapidly. This technique has been experimented with cell suspension apices and somatic embryos of around 20 plant species (Sakai, 1993).

(ii) Encapsulation/dehydration

In this process, explants are encapsulated in alginate beads, cultured for several hours/days in a liquid media with a high sucrose level, then partially desiccated down to water content around 20% (fresh weight basis) and frozen either slowly or rapidly. After slow thawing, beads are transferred to standard medium. Survival rates are high and growth recovery of cryoprserved samples is generally rapid and direct, without callus formation. This technique has been developed for apices of various species from temperate origin such as apple, pear, grape (Dereuddre, 1992, Engelman, 1997).

7.2.2 *In-Situ* Conservation

In situ conservation involves the maintenance of a species in its natural habitat or in the area where it grows naturally. This includes the maintenance of diversity in the farmer's field and orchard, which is termed as 'on farm conservation'.

In situ method is used mostly for conserving the species propagated vegetatively and has non-orthodox seeds and is particularly suitable for the wild and weedy related species of the crops which cannot be established outside their natural habitates. *In situ* conservation is carried out in the form of biosphere or natural reserves, gene sanctuaries, protected forests, national parks, zoological parks etc. Citrus gene sanctuary has already been established in Garo Hills of Meghalaya by conserving the genetic resources of *Citrus* spp in few hundred hectares of forest.

What is *in-situ* conservation?

Conservation of germplasm in natural habitat is known as *in-situ* conservation.

What is gene sanctuary?

Areas of great genetic diversity protected under natural conditions are termed as gene sanctuary.

What is gene bank?

It refers to store of germplasm where genetic resources are maintained in living state.

7.3 International institutes and organizations for germplasm conservation

The important organizations dealing with horticultural germplasm are given below:

(i) International Plant Genetic Resources Institute, Rome, Italy

(ii) International Centre for Tropical Agriculture (CIAT), Ibadan.

(iii) International Potato Centre (CIP), Peru.

(iv) Asian Vegetable Research and Development Centre, Taiwan.

(v) Vavilov Institute of Plant Industry (VIR), Russia.

(vi) Royal Botanical Gardens, Kew, England.

(vii) National Seed Storage Laboratory, Fort Collins, USA.

(viii)National Clonal Germplasm Laboratory, Corvallis, USA.

(ix) National Clonal Germplasm Laboratory, Maimi, USA.

(x) National Clonal Germplasm Laboratory, Hilo, USA.

(xi) National Clonal Germplasm Laboratory, Orlendo, USA.

7.4 National institutes of germplasm conservation

Different government departments dealing with plant genetic resources conservation programes in India namely:

(i) Department of Agricultural Research and Education, including ICAR, Ministry of Agriculture, Government of India.

(ii) Department of Agriculture and Co-operation, Ministry of Agriculture, Government of India.

(iii) Department of Environment and Forestry, Ministry of Environment and Forest, Government of India.

(iv) Department of Commerce, Ministry of Commerce, Government of India.

(v) Ministry of Health and Family Welfare, Govt. of India.

(vi) Ministry of Textile Govt. of India.

(vii) Council of Scientific and Industrial Research, Ministry of Science and Technology (CSIR), Government of India.

(viii)Department of Biotechnology, Ministry of Science and Technology, Govt. of India.

Table 7.1. Most common methods used for germplasm conservation and the corresponding PGR categories (Engels, 1990)

Methods	Predominantly conserved PGR categories by corresponding method
Biosphere reserve	Ecosystem/biodiversity by and large.
Gene sanctuary	Ecosystem (specific)/ wild species genepool.
Nature reserve	Specific habitat/wild and/or weedy species genepool.
On farm conservation (mass reservoirs, bulk hybrid populations)	Agro-ecosystem/landraces.
Botanical garden/arboretum	Wild species, obsolete cultivars, tree crop germplasm.
Field genebank	Wild species, vegetatively propagated crops, tree crop germplasm.
Plant organ storage	Vegetatively propagated crops, mainly in the form of roots, tubers and bulbs
Seed storage	Plant species which produce fertile and orthodox seeds.
Pollen storage	In principle all species which produce long living pollen.
In-vitro storage	Wild and cultivated species which produce recalcitrant or no seeds, vegetatively propagated crops, disease free germplasm as well as orthodox seeds.
Cryopreservation	Germplasm mentioned above which permits cryopreservation.
DNA and gene libraries	Special genetic stocks; in principle applicable for all germplasm.

Chapter 8

DNA Fingerprinting of Crop Plants

8.1 Introduction

Genetic resources must be systematically evaluated to assess genetic diversity. Genetic diversity can be detected based on morphological traits and DNA fingerprinting. However; morphological traits are highly influenced by the environment. Under such circumstances, DNA fingerprinting is an important approach to identify duplicates in the germplasm collections at gene level. DNA fingerprinting refers to unique DNA sequence variation, which are used for asssing genetic similarity and diversity in plant species. Assessment of plant genetic resources is necessary for their efficient and effective management as well as utilization. The establishment of identity of a genotype has assumed greater significance in the context of intellectual property right and trade agreements. A number of molecular tool and procedure are being employed to establish DNA fingerprinting profile.

8.2 Procedure of DNA fingerprinting

The DNA fingerprinting requires following six steps (David, 1999):

8.2.1 Isolation of DNA

DNA must be isolated from the cells of the plant. Only small amount of DNA is needed.

8.2.2 Cutting, sizing and sorting the DNA

The DNA is cut into fragment of different lenghts with the help of restriction enzyme at specific nucleotide sequence (restriction site). The fragments are then separated into bands by electrophoresis. Electrophoresis of the DNA's equivalent of sieving sand through progressively finer mesh screens to determine particle size distribution.

8.2.3 Blotting of DNA

The DNA pieces are transferrd to a nylon or nitrocellulose membrane filter by placing the membrane on the agarose gel.

8.2.4 Probing

The fingerprinting is generated by adding tagged probes to the nylon sheet. DNA probe binds to specific DNA sequence on the nylon membrane.

8.2.5 DNA fingerprint

The excess probe material is washed away leaving the unique DNA based pattern. The radioactive DNA pattern is transferred to X-rays film by direct exposure. When developed, the resultant visible pattern is the fingerprint.

8.3. Application of DNA fingerprinting in germplasm

The cost and genetic stability are important considerations of genetic conservation. DNA fingerprinting reduces the cost of maintenance of germplasm in the genebanks and also ascertains the genetic stability, since it indicates the variability/similarity at DNA level.

A large number of duplicate even multiplicate samples get collected while sampling the germplasm. Similarly one cultivar is conserved in a field genebank by different names. With the application of DNA fingerprinting such duplication of accessions can be avoided in gene banks reducing the cost of conservation.

Chapter 9

Trade Related Intellecutal Property Rights (TRIPs) and IPRs

9.1 Introduction

The importance of intellectual property in India is well established at statutory, administrative and judicial levels. India ratified the agreement establishing the World Trade Organisation (WTO). This has an Agreement on Trade Related Aspects of Intellectual Property Rights (TRIPs) which came into force from 1st January 1995. Intellectual property rights (IPRs) are monopoly right to provide protection for inventions demonstrating the key characteristics of novelty, non obviousness, utility and sufficient disclosure.

9.2 Forms of IPRs

The Agreement provides for norms and standards in respect of following areas of intellectual property:

9.2.1 Patents

Patent is an important area under TRIPs agreement. A patent is a protection granted by the government to an inventor to protect his/her invention for a limited period. Thus, patent rights give the owner the exclusive right to make the invention, use the invention, sell the invention and distribute the invention generally for a period of 20 years. The term of the patent was originally 14 years. This was amended under TRIPs obligations to 20 years. Patents may be granted for all types of processes and products, including those related to the primary sector of production, namely agriculture, fishing or mining, etc. Patenting of traditional knowledge of the people by a person is called biopiracy.

9.2.2 Copyrights and related rights

Copyright protection is provided to the authors of original and scientific works. It protects artistic and literary works and offer strict protection for form and substance only. The owners of copyright can generally prevent the unauthorized reproduction, distribution, sale and adaptation of an original works. Protection generally lasts for the life of the author.

9.2.3 Trade Marks

Trademarks means any symbol, logo or name registered by a manufacturer or merchant to identify goods and service. Protection is usually granted for ten years and is renewable as long as the trademarks continue to be used.

9.2.4 Geographical Indicators

These are signs or expression used to indicate that a product or service originates from a particular territory or a region or locality (Darjeeling Tea and Basmati etc.).

9.2.5 Industrial Designs

It allows the owner of the design to prevent the unauthorized reproduction and distribution of such designs. Industrial designs are created independently that should be new and original. The term of protection generally ranges between 5-15 years including renewal.

9.2.6 Trade secrets

Trade secrets are protected as long as the information is kept secret. A trade secret is any information that can be kept secret to give competitive advantage to the owner over those who do not know it. Secret trading is a common practice for industrial products where the inventors keep a secret of the composition.

9.2.7 Breeders' rights

This is a *sui generis* form of protection conferred on plant varieties that must be novel, distinct, and uniform and stable (DUS). A *sui generis* form of protection introduced for the first time in the USA in 1984. India has also opted a *sui generic* system for the grant of protection to new plant varieties. Exclusive rights, include the sale and distribution of the propagating materials for around 20 years. Unlike patents, breeders' rights permit the use by other breeders of a protected variety as a basis for the development of a new variety (the 'breeders' exemption') and for the re-use by farmers of seeds obtained from their own harvests (the 'farmer's privilege').

Table 9.1. Types of intellectual property rights, subject matter and main areas

Types of intellectual property rights	Subject matter	Main areas
Patents	New, non-obvious, industrially applicable inventions	Chemicals, drugs, plastics, engines, turbines, electronics, industrial control and scientific equipment.
Trademarks	Signs or symbols to identify goods and services	All industries
Copyright	Original works of authorship	Printing, audio, video, motion pictures, software, broadcasting
Breeders' rights	New, stable, homogeneous, distinguishable varieties	food and agriculture industry
Trade secrets	Secret business information	All industries
Industrial designs	Ornamental designs	Automobiles clothing, electronics, etc.
Geographical indications	Geographical origin of goods and services	Spirits, wines, , and other food products

Chapter 10

Vegetable Crops

10.1 Tomato

10.1.1 Introduction

Tomato (*Lycopersicon esculentum*) is one of the most widely grown vegetable in the world. The poupularity of tomato among consumers has made it an important source of vitamins A and C in diets.

10.1.2 Species

The genus *Lycopersicon* designated into the following two subgenera:

10.1.2.1 *Eulycopersicon*: It is characterized by red fruited edible species with carotenoid pigmentation. The species namely *L. esculentum* and *L. pimpinellifolium* are included in this sub genera.

10.1.2.2 *Eriopersicon*: It is characterized by green fruits type with anthocyanin pigmentation. These are *L. pissisi; L. peruvianum;* L. *hirsutum;* L. *glandulosum; L. cheesmani, L. chemelewskii; L. parviflorum, L. chinense.*

10.1.3 Genetic resources resistance/tolerance for biotic stresses

10.1.3.1 Fruit borer: Sabour Prabha, Roma, Sel.152, HS-102, Punjab Kesari, Punjab Chhuhara, S-12, KS-12, *L. hirsutum, L. glabratum.*

10.1.3.2 White fly (*Bemisia tobacci*): *L. peruvianum, L. hirsutum.*

10.1.3.3 Root knot nematode (*Meloidogyne spp*): Sel. 120 (IARI), Hissar Lalit, NRT3, NRT-8, NRT12 (Hissar), Pusa Hybrid 2, Mangla, Arka Vishal, Arka Vardan, TH2312, Pusa 120, Namura, Restino, *L. peruvianum.*

10.1.3.4 Bacterial wilt: Arka Abha, Arka Alok, Swarna Naveen, VC48-1, Shakti, Utkal Pallavi, Utkal Deepti, BWR-1, BWR-5, LE79, PI127805, Pant Bahar, BT-10, BT-1, Sonali, LE 795, B18, B21, Arka Shreshta, Arka Abhijit, Sonali.

10.1.3.5 Fusarium wilt: Pan American (race1), Walter (race2), PI 13448, Roma Columbia, HS110, Pant Bahar, Patriot (EC179883), Marglobe, Mangla, Roma, Rosa.

10.1.3.6 TMV: CL1131-0-013-0-6, CL1131-0-0-43-8-1, CL5915-154, CL143-0-4-13-1, DH1-D.

10.1.3.7 Septoria leaf spot ***(Septoria lycopercisi)*****:** KT4, EC 2750, EC 4555, EC 6993, EC 7785, PI 126448, *L. hirsutum L., L. peruvianum L., L. pimpinnelifolium, L. glandulosum.*

10.1.3.8 Leaf curl (***Bemisia tobacci*****):** H2, H11, H17, H23, H24, H36, H86, H88, Hissar Lalima, Hisar Anmol, *L. hirsutum, L.* glabratum, *L. peruvianum, L. pimpinellifolium, L. hirsutum.*

10.1.3.9 Late blight: Ottawa-30, Ottawa-31, PI-112835, PI-207407, West Virginia-63.

10.1.4 Genetic resources resistance/tolerance for abiotic stresses

10.1.4.1 Low temperature: Fire Ball, Red Kloud, Girl, Immuna, Coldset, Pusa Sheetal, Tempo, S699, Balkan, AC142, Ostam Aniskiji.

10.1.4.2 High temperature: Philippine, Punjab Tropic, EC 130046, EC 162935, EC 321425, EC321426, EC321898, CL5915-154, CL143-0-4-13-1, DH1-D, *L. cheesmanii, L. pimpenneliifolium.*

10.1.4.3 Salt: Sabour Suphala, P-23, Pusa Ruby, Best of All, *L. cheesmanii.*

10.1.4.4 Drought: EC130042, Sel. 28, IIHR 14-1, IIHR 146-2, IIHR 383, IIHR 553, IIHR 555, *L. pimpinellifolium* (PI205009).

10.1.4.5 Moisture: Shakti, BT-10, Arka Abha, Arka Alok.

10.1.4.6 Aluminium: Pant Bahar.

10.1.5 Genetic resources for yield and quality traits

10.1.5.1 High yield : Sioux, Roma, La Bonita, Money Maker, Punjab Kesari, Punjab Chhuhara, Arka Vikash, Arka Saurabh, CO3, EC 106285, EC 118292, EC 339059, EC 3307827, EC 24147, VC 48- 1, Sonali.

10.1.5.2 Large fruit: EC 163708, EC 41340, EC 267728, EC 367856.

10.1.5.3 Red colour: High Pigment, Crimson.

10.1.5.4 High TSS: A 408, A 276, EC 31767, EC 81841, Red Cherry, *L. cheesmani, L. chmielewskii.*

10.1.5.5 High ascorbic acid: L. *pimpinellifolium, L. peruvianum.*

10.1.5.6 Processing: Naveen, Prima, Arka Ahuti, Arka Ashish, Arka Vikas, Arka Ashish, EC 126676, EC 4708, EC 1154, Pant T3, Pusa Gaurav, Pusa hybrid 2, Roma, Punjab Chhuhara, Pusa Uphar.

10.1.5.7 Rich carotene: Pusa Red Plum.

10.1.5.8 Early maturity: EC 357833, EC 357827, EC 362944, EC 362948, EC 99935.

10.1.5.9 Late maturity: EC 357829, EC 362941, EC 362951, EC 241134.

10.1.5.10 Long distance transport: Punjab Chhuhara, Pusa Uphar, La Bonita.

10.2 Brinjal

10.2.1 Introduction

Brinjal (*Solanum melongena* L) is an important vegetable and is part of the regular Indian vegetarian meals. Meshed brinjal is a very popular dish in North Indians. It has medicinal properties particularly for diabetic patients.

10.2.2 Species

There are 5 species namely, *melongena, coagulans, xanthocarpum, indicum* and *maccani*. These are prickly and diploids.The common wild species are *S. torvum, S. indicum* Linn, *S. insanum, S. surattense* Burm. *S. trilobatum. Solanum melongena* L. can be classified based on fruit shape. These are *S. melongena* var. *esculentus*, (round and egg shaped fruit), *S melongena* var. *serpentum* (long and slender types) *Solanum melongena* var *depressum* L (early and dwarf).

10.2.3 Genetic resources resistance/tolerance for biotic stresses

10.2.3.1 Shoot and fruit borer (*Leucinodes orbonalis*)- PPC, PBR-129-5, Punjab Barsati, ME-12, ARV-2C, SM-17-4, Annamalai, Pant Samrat, Bhagyamati, Aushay, Pusa Purple cluster, *S. gilo, S. anomalum, S. khasianum, S. sisymbrifolium, S. melongena.*

10.2.3.2 Jassids (*Amarasca biguttulla, cestius*)- Annamali, Pant Samrat, Pusa Purple Cluster, H-128, Junagarh Selection 1, Aushy, R-34, S488-2, S-34, S258, *S. sisymbrifolium, S. xanthocarpum, S. nigrum, S. khasianum.*

10.2.3.3 Bacterial wilt (*Pseudomonas solanacearum*): Pant Rituraj, BWR-12, BM-44, Arka Nidhi, Arka Keshav, Arka Neelkanth, Long Purple , SM81, JC1, JC2, Annamalai, Pusa Purple Cluster, Pant Samrat, Utkal Tarini, Gulla, Vijay hybrid, Swarna Abhilamb, *S. melongena* var. *insanum.*

10.2.3.4 Phomopsis blight (*Phomopsis vexans*) : Pant Samrat, Florida Market, Florida Beauty, Pusa Bharav, Pusa Anupam, BDS7/331, EC305069, EC316274, IC9010, IC89883, IC90013, NIC4263, NIC9420, NIC13021, *S. xanthocarpum, S. indicum, S. gilo, S. khasianum, S. torum, S. sisymbriifolium , S. testiculatum.*

10.2.3.5 Little leaf (*Hoshimanas phycittis* Dist.) : Pusa Purple Round, Pusa Purple Cluster, PPC, H-8, S.252-1-1, S.252-1-1, *S.incanum, S.*

sisymbrifolium, S. viarun

10.2.4 Genetic resources for abiotic stresses

10.2.4.1 Low temperature: cv. Senrya.

10.2.4.2 Salt: Black Beauty, R-34, S-3.

10.2.4.3 Drought: K2591, Supreme, Violette Round, *S. microcarpon*.

10.2.4.4 High temperature: *S. melongena*.

10.2.4.5 Frost: *S. insanum*, Black Tarpedo, Long Tom 4, Black Beauty.

10.2.5 Genetic resources for yield and quality traits

10.2.5.1 High yield: Arka Sheel, Arka Navneet, Punjab Bahar, BDS6, BDS62,C287-1, IC8655, IC6784, IC32240, IC34078, IC74204, C.89823, IC90122, IC90905, IC90925, Azad 3 .

10.2.5.2 Green fruits: Green Long, Arka Shirish, Rajendra Baigan.

10.2.5.3 Round fruits: Arka Navneet, Pant Rituraj, HABR-4, IVBR-1.

10.2.5.4 Processing purpose: H-4, P-8.

10.3 Chilli

10.3.1 Introduction

Chilli *(Capsicum annum* L.) is most widely cultivated species in the world. It is mainly used for flavouring and imparting the pungency of cooked vegetables and meal.

10.3.2 Species

Species namely, *C. annum annum*, *C. baccatum* var. *pendulum*, *C. chinense*, *C. frutescence* and *C. pubescens* are cultivated and *C. annum* var. *avicular*, *C. baccatum* var. *baccatum*; *C. cardenasii* and *C. eximium* are wild progenitor.

10.3.3 Genetic Resources resistance/tolerance for biotic stresses

10.3.3.1 Thrips: Caleapin Red, Chamatkar, NP 46A, BG-4, Pusa Jwala, Bonapuri.

10.3.3.2 Mite: Kalyanpur Red, Punjab Lal, LEC-1.

10.3.3.3 Aphid: Kalyanpur Red, X 1068, Pusa Jwala.

10.3.3.4 Red spider: EC175959 to 175966.

10.3.3.5 Leaf curl: Pant C-1, Pant C-2, Pusa Sadabahar, Pusa Jwala, Punjab Lal, Jawahar 218, *C. angulosum*.

10.3.3.6 TMV: Punjab Lal, Pusa Sadabahar, Marold Giant (EC304963), EC332333, *C. chinesis,*

10.3.3.7 Fruit rot: Jawahar 218, Chinese Giant, Yolo Y, Hungarian Yellow Wax, Spartan, Lorai, *C. chinense,* S-27, S41-1.

10.3.3.8 C*ercospora* leaf spot: California Wonder, Hungarian Yellow Wax.

10.3.3.9 Bacterial wilt: Utkal Rashmi, KAU Cluster, Pant C-1, White Kandari.

10.3.3.10 Mosaic: Pusa Sadabahar, Pusa Jwala, Punjab Lal, Puri Red.

10.3.3.11 Antharcnose: Punjab Lal.

10.3.4 Genetic Resources resistance/tolerance for abiotic stresses

10.3.4.1 Salinity: Sel-1, LCA-235, C-4, Musalwadi.

10.3.4.2 High temperature: EC354883.

10.3.4.3 Drought: NIC23979, NIC23837.

10.3.4.4 Moisture stress: Resham Patty.

10.3.5 Genetic resources for yield and quality traits

10.3.5.1 High fruit yield: Punjab Lal, Pusa Sadabahar, Pusa Jwala, Arka Mohini, Bhaskar, BDJ 2254, EC246007, IC18882, IC20901, IC92153, IC92156, IC92250, IC119556, IC147637, NIC19991.

10.3.5.2 Perennial: Pusa Sadabahar.

10.3.5.3 Highly pungent: Pant C-1, Jawahar Mirch-218.

10.3.5.4 Early maturity: EC332333, Azad Mirch.

10.3.5.5 Green fruits: KA-2.

10.3.5.6 Dry fruits: CCH-2.

10.4 Cucurbits

10.4.1 Introduction

Cucurbits form an important and a big group of vegetable crops cultivated extensively in the India.

10.4.2 Species

Ten genera contain the species namely, *Cucurbita moschata* (pumpkin), *Cucurbita maxima* (winter squash), *Cucurbita pepo* (summer squash), *Cucumis sativus* (cucumber), *Cucumis melo* (muskmelon), *Cucumis melo* var. *momordica* (snapmelon), *Cucumis melo* var. *utilissium* (longmelon), *Cucumis hardwickii* (related species), *Cucumis hystrix* (related species), *Cucumis setosus* (related species), *Cucumis trigonus* (related species), *Luffa cylindrica* (spongegourd), *Luffa acutangula* (ridgegourd), *Luffa hermaphrodita* (related species), *Luffa groveolons* (related species), *Luffa echinata* (related species), *Lageneria siceraria* (bottlegourd), *Citrullus lanatus* (watermelon), *Citrulus colocynthis* (related species), *Citrullus ecirrhosus* (related species), *Cirullus nandinianus* (related species), *Tricosanthes anguina* (snakegourd), *Tricosanthes dioica* (parwal), *Momordica cochinchinensis* (sweetgourd), *Momordica charantia* (bittergourd), *Momordica dioica* (kakrol), *Benincasa hispida* (ashgourd), *Coccinia grandis* (vegetable), *Schium edule* (chayote).

10.4.3 Genetic resources resistance/tolerance for biotic stresses

10.4.3.1 *Cucumis melo* (Muskmelon)

10.4.3.1.1 Aphid: EC350602.

10.4.3.1.2 Water melon mosaic virus: EC350603.

10.4.3.1.3 Watermelon mosaic virus (WMV) and Powdery mildew: EC350604.

10.4.3.1.4 Fusarium: EC382734.

10.4.3.1.5 Downey mildew: Punjab Rasila, Hisar Saras, EC399917, EC399924-31.

10.4.3.1.6 Powdery mildew: Punjab Rasila, EC399934, Arka Rajhans.

10.4.3.2 *Cucumis sativus*

10.4.3.2.1 Scab and powdery mildew: EC320556.

10.4.3.2.2 Multiple diseases: EC329300.

10.4.3.2.3 Anthracnose: EC398968-70.

10.4.3.2.4 Fruit rot: EC398971-73.

10.4.3.2.5 Downey mildew: EC398974-90.

10.4.3.2.6 Downey mildew, powdery mildew and anthracnose: Poinsett.

10.4.3.2.7 Leaf spot: EC398991-399007.

10.4.3.3 *Citrullus lanatus* (Watermelon)

10.4.3.3.1 Wilt: EC333659-69.

10.4.3.3.2 Fusarium and anthracnose: EC382753, Charleston Grey.

10.4.3.3.3 Fusarium: EC383809.

10.4.3.3.4 Anthracnose: Africon Citron 8, Charleston Grey, Congo, Crimson Sweet, Jubilee.

10.4.3.3.5 Powdery and downey mildew: Arka Manik.

10.4.4 Genetic resources resistance /tolerance to abiotic stresses

10.4.4.1 *Cucumis melo*

10.4.4.1.1 Heat: EC342751-54.

10.4.4.2 *Cucurbita foetidssima*

10.4.4.2.1 Drought: EC442706-442713.

10.4.4.3 *Tricosanthes anguina* (snakegourd)

10.4.4.3.1 Drought: PKM 1.

10.4.5 Genetic resources for yield

10.4.5.1 High yield

10.4.5.1.1 *Cucurbita moschata* (pumpkin): Pusa Alankar, Pusa Vikash, IC90512, IC92499, IC92533, IC92571, NIC10170, NIC10174, Arka Chandan, Pusa hybrid 1.

10.4.5.1.2 *Citrullus lanatus* (watermelon): Arka Jyoti, Arka Manik, EC28435, K3228, OC92910.

10.4.5.1.3 *Cucumis melo* (muskmelon): Hara Madhu, Pusa Sharbati.

10.4.5.1.4 *Cucumis sativus* (Cucumber): IC203838, EC237658, VJ/98-176, VJ/98-151, Straight Eight.

10.4.5.1.5 *Luffa cylindrica* (spongegourd): NIC23288, NIC23291, NIC23292.

10.4.5.1.6 *Ligenaria siceraria* (bottlegourd): IC29404, IC92429, IC110389.

10.5 Cole crops

10.5.1 Introduction

Among the cole crops in India, cauliflower and cabbage are extensively grown all over the country while remaining grown in certain areas.

10.5.2 Species

Cole groups comprise the cultivated botanical species namely, Cabbage (*Brassica oleracea* L. var. *capitata*), Portuguese cabbage (*Brassica oleracea* L. var. *costata)*; Tree cabbage (*Brassica oleracea* L. var. *palmifolia*), Savoy cabbage (*Brassica oleracea* L. var. *sabauda*), Cauliflower (*Brassica oleracea* L. var. *botrytis),* Knol khol (*Brassica oleracea* L. var. *gongylodes),* Brussels sprout (*Brassica oleracea* L. var. *gemmifera),* Broccoli, calabrese (*Brassica oleracea* L.var. *italica),* Kale, Collard & tree Kale (*Brassica oleracea* L. var. *acephapa),* Marrow stem kale (*Brassica oleracea* L. *var. medullosa), Chinese kale (Brassica Oleracea* L. *var. alboglabra), Collards* (*Brassica oleracea* L. *var. sabellica),* Broccoli (*Brassica oleracea* L. *var. selencia).*

10.5.3 Genetic resources resistance/tolerance for biotic stresses

10.5.3.1 Black rot (*Xanthomonas campestris)*

10.5.3.1.1 Cauliflower: Pusa Snowball, K-1, Pusa Shubhra, Selection line 6-1-2-1.

10.5.3.1.2 Cabbage: Pusa Mukta (Sel. 8), Green Land, Hammar, EC10109, Spitz Kool, EC24855.

10.5.3.2 Black leg (*Phoma lingam)*

10.5.3.2.1 Cabbage: Pusa Drumhead, Pusa Mukta, Pride of India.

10.5.3.2.2 Cauliflower: Pusa Shubhra.

10.5.4 Genetic resources resistance/tolerance for abiotic stresses

10.5.4.1 Cabbage

10.5.4.1.1 Heat: Pusa Ageti.

10.5.4.1.2 Aluminium: Pusa Drumhead.

10.5.5 Genetic resources for quality traits

10.5.5.1 Early varieties

10.5.5.1.1 Cauliflower: Early Kunwari, Pusa Early Synthetic, Pusa Deepali, Pusa Meghna, Pusa Katki, IVREC-2, SYCFH-203, Improved Japanese, Pant Gobhi 3.

10.5.5.1.2 Cabbage: Cophenhagen Market, Golden Acre, Pride of India.

10.5.5.1.3 Knol khol: Early White Vienna, White Vienna.

10.5.5.2 Quality curds

10.5.5.2.1 Cauliflower: Pusa Snowball, K-1.

10.6 Okra

10.6.1 Introduction

Okra (*Abelmoschus esculentus* L.) is locally known as bhindi. It is one of the most important summer vegetable crops in India. It is rich source of vitamins Ca, K and other mineral matter.

10.6.2 Species

There are six species *viz., Abelmoschus moschatus* Medikus, *A. manihot* L., *A. esculentus* (L) *Moench, A. ficulneus* (L) *Wt and Art. Ex Wt., A. crinitus* Wall, *A. angulosus* Wall ex Wt. and Art. The former 3 species consist of wild and cultivated forms and the latter three species are wild forms only.

10.6.3 Genetic resources resistance/tolerance for biotic stresses

10.6.3.1 Fruit and shoot borer (*Earias sp*): Pusa Sawani, Long Green, Red Wonder-1, Wonderful Pink, A.E.3, A.E. 22, Pusa A4, *A. manihot.*

10.6.3.2 Aphid: A.E. 11, A.E. 17, Pusa Sawani.

10.6.3.3 Jassids (*Amrasca biguttuli biguttula*): Early Long Green, Pusa Sawani, Punjab Padmini, Pusa A4, Varsha Uphar, Crimson Smoth Long, IHR-21, IC17194, IC8899, AE-22, *A. manihot ssp. manihot, A. moschatus, A. tuberculatus.*

10.6.3.4 Fusariums wilt (*Fusarium oxysporium*): Pusa Sawani, Pusa Makhmali, *A. manihot.*

10.6.3.5 Damping off (*Rhizoctonia solani*): Red Ghana, IC- 17252, IC 12096.

10.6.3.6 Powdery mildew (*Erysiphe cichoracearum*): EC 32598, IC 8248, Nigeria, *A. tetraphyllus, A. manihot, A. moschatus, A. angulosus.*

10.6.3.7 *Cercospora* blight (*Cercospora abelmoschi*) and alternaria blight: EC305616, EC 32598, Round Selection, IC 10238, IC8248, IC1542, EC32598, EC41292, *A. crinitus, A. angulosus, A. moschatus.*

10.6.3.8 Yellow vein mosaic virus: Parbhani Kranti, Azad Kranti, Punjab Padmani, Varsha Uphar, P-7, Arka Anamika (Sel. 10), Hissar Unnat, Arka Abhay (Sel.4), Hisar Barsati, Utkal Gaurav, EC329375, NIC3322, NIC3325, NIC6308, NIC9303 A, NIC9408, IIHR-4, Sel-2, Kashi Vibhuti, Kashi Pragati, Kashi Bhairav, Kashi Satdheri, Shitla

Jyoti, NDO-10, HRB-107-4, TN hybrid 8, *A. manihot, A. tetraphyllus*

10.6.3.9 Root knot nematode: Long Green Smooth, Harichikni, Kanki Local Green, Vaishali Badhu.

10.6.4 Genetic resources resistance/tolerance for abiotic stresses

10.6.4.1 Low temperature and frost: *A. angulosus.*

10.6.4.2 Salinity: Pusa Sawni.

10.6.4.3 Aluminium: Parbhani Kranti.

10.6.5 Genetic resources for yield

10.6.5.1 High yield: Pusa Sawani, Punjab Padmani, Arka Anamika, Parbhani Kranti, HBH-142 (hybrid), SOH-152 (hybrid).

10.7 Carrot

10.7.1 Introduction

Carrot (*Daucus carota*) is a very popular vegetable in India. It is rich in carotene, vitamin A and contains appreciable quantities of thiamine and riboflavin.

10.7.2 Species

Carrot has two cultivated types, *i.e.* Asiatic (*D. carota* subsp. *sativas)* and European (D. *carota* subsp. *carota).* Asiatic type is native to Afganistan, Asia and Africa whereas European type has been cultivated in the Mediterranean region from early times.

10.7.3 Genetic resources resistance/tolerance for biotic stress

10.7.3.1 Alternara blight or late blight: PI261648, PI 226043, Imperial Long Scarlet.

10.7.4 Genetic resources for abiotic stresses

10.7.4.1 Low temperature: Nantes, Zeno.

10.7.4.2 High temperature: No. 29, Pusa Kesar, Sel- 233.

10.7.5 Genetic resources for yield and quality traits

10.7.5.1 High yield: Sel-233, No. 29, Pusa Yamdagni, Golden Ball.

10.7.5.2 High carotene: Pusa Yamdagni, Pusa Kesar, EC198258.

10.7.5.3 Canning: Chantenay, Nantes, Amsterdamer Trib.

10.7.5.4 Shelf life: American Beauty.

10.8 Potato

10.8.1 Introductions

Potato is one of the major food crops of the world. The main ingrediatns of potato are sugar, starch, mineral, vitamins and protein.

10.8.2 Species

It has two subspecies viz., *tuberosum* and *andigena*. The former species adapted to long days and later adapted to short days. The primitive cultivated species are *S.phureja, S. chaucha, S. stenotomum, S. gonicalyx, S. curtilobum, S. juzepczukii* and *S. ajanhuiri.*

10.8.3 Genetic resources resistance/tolerance for biotic stresses

10.8.3.1 Late blight (*Phytopthora infestans*): CP1673, CP 1690, CP2067, Kufri Kuber, CP2407, CP3190, CP3250, CP3276, CP1385, CP1396, CP1436, CP1470, CP1924, CP3174, *S. demissum, S. microdontum* Bitter, *S. sucrense.*

10.8.3.2 Wart (*Synchytrium endobioticum*): CP2291, CP2307, CP2308, CP2322, CP2371, *S. chacoense* Bitter, *S. spegazzinii.*

10.8.3.3 Common scab (*Streptomyces scabies*): CP2213 to CP2263.

10.8.3.4 Cyst nematode (*Globodera pallida & G. rostochinensis*): CPI1664, CP1495, CP1663, Royal Kidney, *C. chacoense, S. sparsipilum.*

10.8.3.5 Root knot nematodes (*Meloidogyne incognita*): *S. microdontum, S. vernei* Bitter & Wittm

10.8.3.6 Leaf hopper (*Amarasca devastans*): CP1558, Sturna, CPI3152, CP3189

10.8.3.7 Potato tuber moth: (*Phthorimaea operculella*): JEX/ A163, JEX/ A250.

10.8.3.8 Stem necrosis: Kufri Sinduri, Hybrids PS/ F220, JN-189.

10.8.4 Genetic resources resistance/tolerance for abiotic stresses

10.8.4.1 Heat: CP2150, CP2118, CO2109, CP2108.

10.8.4.2 Mineral nutrition: JEX/ A Nos. 17, 40, 79, 163, 183, 196, 421, 453.

10.8.4.3 Soil salinity: CP1532 and CP2061, JE808, Kufri Chamatkar, CP2059, JE-303, Kufri Sinduri.

10.8.5 Genetic resources for quality traits

10.8.5.1 Protein: CP2401, CP 1929, CP2142, CP2334, CP2384, CP2401, CP2407, CP2416.

10.8.5.2 Vitamin C: JEX/A Nos.: 63, 68,208, 218, 426, and 483.

10.9 Sweet Potato

10.9.1 Introduction

Sweet potato is one of the important staple food crops of the tropics for their edible tubers. The main ingrediants of sweet potato are water, carbohydrates, protein and fat and also contains carotin and provitamin A.

10.9.2 Species

Important wild species are *Ipomea alba, I. angulata, I. coccinea, I. dumetorum, I. indica, I. labata, I. palmate, I. parasitica, I. piurensis, I. purpurea, I. squamosa.*

10.9.3 Genetic resources resistance/tolerance for biotic stresses

10.9.3.1 Sweet potato weevil: TIS-2532, TIS-3053, TIS-3017, TIS-9365, Rajendra Sakarkand-5, H-42, Rajendra Sakarkand-43.

10.9.3.2 Fungal diseases (blackrot): B-5941, B-5944, B-5999.

10.9.3.3 Viral disease (feathery mouth complex): TIS-2498, TIS-9265 (IITA).

10.9.4 Genetic resources resistance/tolerance for abiotic stress

10.9.4.1 Drought: H-268, Varsha, Sree Vardhini.

10.9.5 Genetic resources for yield and quality traits

10.9.5.1 High yield: Pusa Saffaid, V-2., No. 4004, Kal Megh, Rajendra Sakarkand-35.

10.9.5.2 Rich in carotene: Pusa Sunehri, V.8, Sree Vardhini, Kiran, Gouri.

10.9.5.3 Nonfibrous tuber: Tie Shen Tun, H.42.

10.9.5.4 High keeping quality: Pusa Lal.

10.9.5.5 Short duration: CO2.

10.9.5.6 Early maturity: Sree Nandini, Sree Vardhini, CO2, CO3, Samrat, Kiran, Sree Bhadra.

10.9.5.7 Excellent cooking quality: H41, H42, Kanjagad, Sree Rethna, Sree Bhadra, Sankar.

10.10 Onion

10.10.1 Introductions

Onion (*Allium cepa* L.) is one of the most important vegetable crops grown throughout the world. Onion can be used as salad, cooked all curries. It is also used for making pickles.

10.10.2 Species

The onion has three botanical varieties namely, *Allium cepa* var *cepa, Allium cepa* var. *aggregatum* and *Allium cepa* var. *proliforum.*

10.10.3 Genetic resources resistance/tolerance for biotic stresses

10.10.3.1 Purple blotch: Arka Nihar, Arka Kalyan, IIHR56-1, VL-67, Pusa Ratnar, N-2-4-1, Pusa Red, Punjab Selection-3, $CO\text{-}_2$.

10.10.3.2 Alternaria blight: N-2-4-1.

10.10.3.3 Basal rot: Poona Red, Patna Red, Bellari Red, IHR-41, N-257-9-1.

10.10.3.4 Thrips: Pusa Ratnar, Kalyanpur Red Round, Udaipur-103, Bombay White, Spanish White, Hissar-2, Arka Suryamukhy, N-2-4-1, CO3 and CO4.

10.10.4 Genetic resources resistance/tolerance for abiotic stresses

10.10.4.1 Salinity: Hissar-2, Punjab Selection, Udaipur-102, Bombay Red, Pusa Ratnar.

10.10.4.2 Lodging: MDU-1.

10.10.5 Genetic resources for quality traits

10.10.5.1 Attractive colour: IC34540, IC42900, IC47974, IC48000, IC 48205, IC488996.

10.10.5.2 High TSS in bulbs: IC35758, IC40051, IC42854, IC47969, IC48183, IC48503, IC48573.

10.10.5.3 High tillering: IC48230, IC48236, IC49207, IC489968.

10.10.5.4 Early maturity (seed): IC3444, IC48025, IC48051, IC34448045.

10.10.5.5 Good keeping quality: MDU 1, Arka Pitambar.

10.11 Garlic

10.11.1 Introductions

Garlic (*Allium sativum* L.) is one of the major bulb crops grown throughout India. It is used as a spice or condiment throughout India. Garlic is considered as a rich source of carbohydrates, proteins and phosphorus.

10.11.2 Genetic resources resistance/tolerance for biotic stresses

10.11.2.1 Garlic mosaic virus: IC35325, IC49373, IC49384.

10.11.2.2 Purple blotch and mold: EC328492, EC328494, EC328501.

10.11.2.3 Thrips: G-2, G6, G19, G20, G31, G43, G84.

10.11.3 Genetic resources resistance/tolerance for abiotic stress

10.11.3.1 Salinity: HG-6.

10.11.4 Genetic resources for yield and quality traits

10.11.4.1 High bulb yield: IC39382, IC49373, IC493882.

10.11.4.2 Bold clove: EC97537, EC172362, EC182916.

10.11.4.3 High foliage and high number of cloves: EC244954, EC244957, EC244972, IC32320, IC49373.

10.12 Pea

10.12.1 Introductions

Garden pea is a most important vegetable crop. It is consumed as fresh vegetable or dry seeds. Peas are nutritive and are rich source of protein, carbohydrate and vitamins.

10.12.2 Species

There are two subspecies, viz., *Pisum arvense* has coloured flowers known as field pea whereas Pisum *sativum* has white flowered known as garden pea or sweet pea.

10.12.3 Genetic resources resistance/tolerance for biotic stresses

10.12.3.1 Aphid and weevil: PI244241, LMR-4, LMR -10, LMR-20.

10.12.3.2 Stem fly: Sophia.

10.12.3.3 Leaf minor: LMR-4, LMR-10, LMR-20, JP179, JP747, Arkel, Kashi Nandini.

10.12.3.4 Powdery blight: PRS-4, T-10, P 388, JP 83, T 6587, 6588, JP-54, T-56, P-185, Sel. 1, EC 3386, Lines P 185, 6585, 6587, Giant Sugar, American Wonder, PMR lines, Arka Ajit, Pant Matar 2, Kashi Mukti, KTP-8.

10.12.3.5 Fusarium wilt: Grey Badger, Alaska, Early Giant, JM-1, JM-2, Sel123-3-2, GC468, JP501/2.

10.12.3.6 Pea mosaic: American Wonder, Little Marvel.

10.12.3.7 Pea rust: JP-4.

10.12.3.8 *Ascochyta* blight: Austrian Winter, Kinnauri.

10.12.4 Genetic resources resistance/tolerance for abiotic stresses

10.12.4.1 Salinity: P-23, New Line Perfection, Market Prize.

10.12.4.2 Frost: Multifreezer.

10.12.4.3 Heat and drought: Early Badger, Alsaka.

10.12.5 Genetic resources for yield and quality traits

10.12.5.1 High yield: Perfection New line, Hisar Lalit, Matar Ageti 6, Kashi Mukti, Kashi Shakti.

10.12.5.2 Early varieties: Arkel, Pant Uphar, Hisar Harit, Jawahar Matar-4, Alaska, Pant Sabzi Matar, Ageti Matar-7, Pant Matar 2, Asauji, Early Superb, Meteor, Early Badger, Little Marvel, Matar Ageti 6, Kashi Nandini, Kashi Udai, Kashi Mukti, VRP-5.

10.12.5.3 Canning: Arka Ajit.

10.13 French bean

10.13.1 Introduction

French bean is one of the most important leguminous vegetable. It is grown for pods as well as seeds. It is extensively grown as green vegetable in Hill areas.

10.13.2 Species

There are four cultivated species of *Phaseolus* namely, *P. vulgaris, P. coccineus, P. lunatus* and *P. acutifolius* var. *latifolia.*

10.13.3 Genetic resources resistance/tolerance for biotic stresses

10.13.3.1 Bean common mosaic virus: Pant Anupama, RH13, Viva, ARS 6B65, ARS 5 BP-7, Sea Farer, Robust, Viva Roza, Gloria, NW 410, NW 590, Swan valley, Turkish Brown.

10.13.3.2 Anthracnose: Cornell 49-242, Wells Red Kidney, Sea Farer, Ooty 1, RH13, Nairobi Acc. No. 16, 84, 86, Phule Surekha.

10.13.3.3 Angular leaf spot: Pant Anupama, Lakshmi, SVM-1, Ooty 1.

10.13.3.4 Powdery mildew: Pusa Parvati, Contender, Hungarian Yellow, Pusa Parvati, Hungarian Yellow, Lady Washington, Pinto, Ma carrao, Favorites.

10.13.3.5 Fusarium yellows: Early Gallatin, Pintado.

10.13.3.6 Rhizoctonia wilt: Manoa Wonder, Cornell 2114-12, PI-165426.

10.13.3.7 Pod borer: Ooty 1.

10.13.3.8 Root knot nematode: Albama No1, Albama No.2.

10.13.4 Genetic resources resistance/tolerance for abiotic stress

10.13.4.1 Heat: 5 BP 7, Haibushi.

10.13.5 Genetic resources for yield and quality traits

10.13.5.1 High yield: Contender, Giant Stringless, Big Ben Red, VL Boni, Pusa Parvati

10.13.5.2 Good pod quality: IVFB-1.

Chapter 11

Fruit Crops

11.1 Mango

11.1.1 Introduction

The mango is one of the oldest cultivated fruit crops. It is widely grown in India for atleast 400 years and is also popular in south-east Asian countries.

11.1.2 Species

The following species are given below:

11.1.2.1 Wild Species

M. gedebe Miq, *M. caloneura* Kurz, *M. cochinchinensis* engier, *M. andamanica* King, *M. parvifolia* Boeri & Koorders, *M. pedicellata* Kosterm, *M. dewildei* Kosterm, *M.quadrifida* Jack, *M. rubrocostata* Kosterm, *M. sulavesiana* Kosterm, *M. linearifolia* (Mukherji) Kosterm, *M. sumbawaensis* Koterm, *M. sylvatica* Roxb., *M. swintoniodes* Kosterm, *M. zeylanica* (BI.) Hooker f., *M. blommesteinii* Kosterm, *M. macrocarpa* BI, *M. decandra* Ding Hou, *M. lagenifera* Griff, *M. persiciformis* WU & Ming, *M.bullata* Kosterm, *M.transversalis* Kosterm.

11.1.2.2 Cultivated species

M. gracilipes Hooker f, *M. minutifolia* Evard, *M. griffithii* Hooker f, *M. microphylla* Griff. Ex Hooker f, *M. altissima* Blanco, *M. applanata kosterm, M.oblongifolia Hooker f, M. orophila* Kosterm, *M.casturi* Kosterm, *M. rubropetala* Kosterm, *M. Indica* L., *M. laurina* BI, Kochummen, *M.timorensis* BI, M. minor BI., *M. torquendai kosterm, M. leschenaultii* Marchand, *M. caesia* Jack, *M.odorata* Griff, *M.pajong koserm, M. kemanga* BI, *M.bompardii* Kosterm, *M. lelijiwa* Kosterm.

11.1.3 Genetic resources resistance/tolerance for biotic stresses

11.1.3.1 Hoppers, tip borers and seed borers: *M. altissima.*

11.1.3.2 Malformation: Bhadauran.

11.1.3.3 Powdery mildew: Neelum, Totapuri, Zardalu.

11.1.3.4 Anthracnose: Tommy Atkins, Parish, Fairchild, *M. laurina, M. minor.*

11.1.3.5 Bacterial canker: Bombay, Ambrapali.

11.1.4 Genetic resources resistance/tolerance for abiotic stresses

11.1.4.1 Cold: *M. orophila.*

11.1.4.2 Waterlogged condition: *M. gedebe, M. decandra, M. inocarpoides.*

11.1.4.3 Calcareous soil: M13.

11.1.5 Genetic resources for yield and quality traits

11.1.5.1 High yield: Totapuri, Neelum, Sindhu, Hybrid-1, Hybrid 2, Hybrid-3.

11.1.5.2 Dwarfness: Creeping, Kalapadi, Belkhas, Parikhas.

11.1.5.3 High Sex ratio: Dashehari, Langra, Neelum, Amrapali, Alphonso, Raspuri.

11.1.5.4 Self Incompatibility: Dashehari, Langra, Chausa, Bombay Green.

11.1.5.5 Big fruit size: Banganapalli, Janardan Pasand, Kensington, Fazali, Langra, Malgoa, Mallika.

11.1.5.6 Red peel colour: Janardan Pasand, Sinduria, Benazir, Fernandin, Suvarna Rekha, Vanraj, Kensington, Tommy Atkins, Sensation, Irwin, Arka Puneet, Nazuk Badan, Raspuri.

11.1.5.7 High pulp content: Sindhu, Sahib Pasand, Amrapalli, Sunderja, Mallika, Bhutto, Bombay, Fazli, Khudadad, Chausa, Sora, Vanraj, Langra, Dashehari, Pairi, Arka Aruna, Jawahar.

11.1.5.8 High TSS content: Sahib Pasand, Doodh Peda, Dashehari, Langra, Arka Puneet, Mallika, Lazzat Baksh, Amrapalli, Sakkar China, Gopal Bhog, Rani Pasand, Sunderja, Neelphonso, Neeleshwari.

11.1.5.9 Good keeping Quality: Alphonso, Totapuri, Sundershah, Dashehari, Banganpalli, Kesar, Rataul, Arka Anmol, Arka Puneet, Mallika, Fazli, Gulab Khas, Himsagar, Kishenbhog, Mankurad, Neelum.

11.1.5.10 Regular bearing: Neelum, Kalapadi, Totapuri, Arka Anmol, Arka Neelkiran, Kesar, Rumani, Arka Aruna, PKM-1, PKM-2, Mahmud Bahar, Prabha Shankar, Sabori, Jawahar, Neeludin, Neeleshan, No.2/2, Mallika, Ruchi, Neeleshan Gujrat, Neeleshwari,

Sai Sugandh.

11.1.5.11 Early maturity: Gaur Jeet, Mithua, Zardalu, Lazzat Baksh, Guruvarm, Chandrakaran, Panakalu, Olur.

11.1.5.12 Late maturity: Katki, Bathui, Fazli, Neelum, Arka Anmol, Chausa, Fakhr-us-Samar, Puttu, Sora, Sultanus-Samar, Azam-us-Samar, Totapuri.

11.1.5.13 Processing (pulp/juice):Alphonso, Rani Pasand, Sahid Pasand, Pairi, Baganpalli, Totapuri, Hyder Saheb, Pedda Rasam, Sukul, Allumpur Baneshhan, Neelum, Chausa, Alif Laila, Prabha Shankar, Kesar, Arka Anmol, Arka Puneet.

11.1.5.14 Pickle and Chutney: Amlette, Chandrakaran, Karanjio, Ram Kela and other mango varieties at unripe stage.

11.1.5.15 Freestones: *M. similes* (good parent for breeding).

11.2 Banana

11.2.1 Introduction

Banana is a rich source of several minerals such as calcium, magnesium, potassium, phosphorous and has several medicinal properties.

11.2.2 Species

Cultivated and wild species are given below:

11.2.2.1 Cultivated: *Musaceae* with two genera, *Musa* and *Ensete* with forty species. Only *Musa* has commercial importance with a number of species, *viz., M. flavicarpa, M. itinerans, M. basjoo, M. schizocarpa* and *M. acuminata.*

11.2.2.2 Wild: *Musa acuminata* ssp. *burmannica, M. bulbisiana, M. nagensium Prain, M. sikimensis* Kurz, *M cheesmani* Simmonds, *M. ornate* Roxb, *M. laterita,* Chees, *M. velutina Wendl* Ex Drude and *M. sanguinea*

11.2.3 Genetic resources resistance/tolerance for biotic stresses

11.2.3.1 Burrowing nematode: Kadali, Octoman, Thaen Kunnan, Ney Poovan, Kunnan, Tongat, Pisang Lilin, Dudh Sagar, Chini Champa, Anil Vazhai, Manik Champa, H_1.

11.2.3.2 Panama wilt: Basrai, Poovan, Champa, Cavendish group, Mongil, Peyladen, Rajabale, Vamanakeli, Giant Governor, Robusta.

11.2.3.3 Bunchy top: Poovan.

11.2.3.4 Fusarium wilt: BITA-1, BITA2.

11.2.3.5 Leaf spot: H1, H2.

11.2.4 Genetic resources resistance/tolerance for abiotic stresses

11.2.4.1 Drought: Karpuravalli, FHIA-2.

11.2.4.2 Salt: Karpuravalli.

11.2.5 Genetic resources for high and yield quality traits

11.2.5.1 High yield: FHIA-6.

11.2.5.2 Good keeping quality: Hill Banana, Ney Povan.

11.2.5.3 Suitable for long distance transportation: Karpuravalli.

11.2.5.4 High sugar content: H1.

11.2.5.5 Good pulp quality: Goldfinger.

11.3 Citrus

11.3.1 Introduction

Citrus constitutes a major group of fruits that is comprised of orange, mandarin, lime, lemon, grapefruit, pummelo, tangelo, lemonime, kumquat, trifoliate orange, citron, citrange etc. Among these, mandarin (*Citrus reticulata*) followed by sweet orange (*Citrus sinensis*), and acid lime (*Citrus aurantifolia), lime (Citrus limon)* are extensively cultivated in India.

11.3.2 Species

The distribution of various indigenous *Citrus* species in India is given below:

- **Distributed all over the country :** *C. aurantifolia, C. grandis, C. jambhiri, C. karna, C. limon* var. *lomon, C. medica, C. reshni, C. reticulata, C. aurantium, C. semiperflorens.*
- **North-western India :** *C. aurantifolia, C. limettioides, C. limon var. decumana, C. reshni, C. reticulate var. crenatifolia, C. rugulosa, C. karan.*
- **North eastern India :** *C. assamensis, C. karna, C. latipes, C. limon var. decumana, C. limon, C. indica, C. limetta , C. limonia, C. macroptera, C. megaloxycarpa, C. sinesis, C. nobilis, C. inchangensis.*
- **South India :** *C. pennivesiculata, C. medersapatana, C. nakoor.*
- **Central India :** *C. limonia.*

11.3.3 Genetic resources resistance/tolerance for biotic stresses

11.3.3.1 Rootrot: Rangpur lime, Marmalade orange, Troyer citrange, Carrizo citrange.

11.3.3.2 Nematode: Marmalade orange, Sweet lime, Trifoliate orange, Troyer citrange, Carrizo Citrange, Karna Khatta, Nasnaran, *Survernia disticha,* CRH-3, CRH-5, CRH-41 (IIHR, Bangalore).

11.3.3.3 Tristeza virus: Rangpur lime, Marmalade orange, Rough lemon, Cleopatra mandarin, Trifoliate orange, Troyer Citrange, Carrizo Citrange, Nasnaran, Surki mandarin, sweet orange.

11.3.3.4 Xylopsorosis: Rough lemon, Trifoliate orange, Sweet orange, Sour orange, Cleopatara mandarin, grape fruit, Troyer Citrange.

11.3.3.5 Exocortis: Rough lemon, Sweet orange, Sour orange, Cleopatra mandarin, Grape fruit.

11.3.3.6 Gummosis: Trifoliate orange, Sour orange, Rangpur lime, Cleopatra mandarin, Troyer Citrange.

11.3.3.7 Greening: Amilbed, Gajanima, Florida Rough lemon, Karna Khatta.

11.3.4 Genetic resources for abiotic stresses

11.3.4.1 Drought: Rangpur lime, Marmalade orange, Rough lemon, Sour orange.

11.3.4.2 Salt: Rangpur lime, Marmalade orange, Nasnaran, Cleopatra mandarin, Sunki mandarin.

11.3.4.3 Aluminium: Cleoptera mandarin, Rough lemon.

11.3.5 Genetic resources for yield and quality traits

11.3.5.1 Rootstock suitable for good yield: Rangpur lime, Marmalade orange, Rough lemon, Cleopatra mandarin, Sour orange, Sweet orange, Karna Khatta.

11.3.5.2 Rootstock suitable for good quality fruits: Sour orange, Troyer Citrange, Carrizo Citrange, Sweet orange, Nasnaran.

11.4 Guava

11.4.1 Introduction

The guava (*Psidium guajava L.*) is one of the most common fruits in India. It is rice source of ascorbic acid as compared to citrus.

11.4.2 Species

Guavav has several species, like, *P. cattleianum sabine* (strawberry *guava), P. friedrichsthalianum (costa rican guava), P. montanum, P. littorale var. longipes, P. longipes. Some wild species, namely, P. cujavillus, P. molle, P. cattleianum, P. guineese, P. pumilum, P. chinensis, P. acutangulum* are introduced and maintained.

11.4.3 Genetic resources resistance/tolerance for biotic stress

11.4.3.1 Wilt: *P. friedrichsthalianum, P. chinensis, Feijoa sellowiana.*

11.4.4 Genetic resources resistance/tolerance for abiotic stresses

11.4.4.1 Drought: L-49, Allahabad Safeda, Chittidar, Safed Jam, Kohir Safeda.

11.4.4.2 Cold: Allahabad Safeda, *P. cattleianum.*

11.4.5 Genetic resources for quality and yield traits

11.4.5.1 High yield: L-49, Allahabad Safeda, Lalit, Rewa-72, Chittidar, Dhareedar, Arka Mridula, CISH-G4.

11.4.5.2 Pink skin colour: Apple colour, CISH-G_{-2}, CISH-G_{-1}, Hybrid-16-1 (Apple colour x Allahabad Safeda), Allahabad Surkha.

11.4.5.3 Pink flesh colour: Ruby, Blitch, Anakapalli, Red Fleshed, Super Acid, Florida Seedling, Hybrid Ruby Supreme, Purtagal, China Surkha, Kothred, Patillo, Beaumont.

11.4.5.4 Pink skin with pink flesh colour: Lal Sahiba, Sangam.

11.4.5.5 Juice making: Safed Jam, Kohir Safeda.

11.4.5.6 Soft seed: Hybrid-1 (Allahabad x Seedless), Hybrid-16-1 (Apple colour x Allahabad Safeda), Safed Jam, Kohir Safeda, Chittidar.

11.4.5.7 High TSS: Allahabad Safeda, Chittidar, CISH-G-1, China White, Barainpur, Kothred, Burafkhana.

11.4.5.8 Dwarf rootstock: *P. chinensis,* Aneuploid-82, and other Aneuploids, Hybrid-1.

11.4. 5. 9 Good keeping quality: Hārijha, Arka Mridula, Arka Amulya, Hybrid 16-1.

11.5 Grape

11.5.1 Introduction

The grape is the most important deciduous crop. It contains various organic and inorganic components of fresh juice.

11.5.2 Species

Genus *Vitis* is divided into two sections: *Euvitis* and *Muscadinia*. The important grape species included in *Euvitis are, V. vinifera, V. amurensis* and *V. rupestris.*

11.5.3 Genetic resources resistance/tolerance for biotic stresses

11.5.3.1 Phylloxera vitifoliae: *V. rupestris,* Harmony, Dog Ridge, *V. riparia,* St. George, 99-R, *V. berlandeiri, V. monticola, V. aestivalis.*

11.5.3.2 Aphids: *V. cinerea.*

11.5.3.3 Thrips: Karachi, Fakdi.

11.5.3.4 Nematode: Dog Ridge, Salt Creek, 1613 (Solonis x V. *riparia*), Teleki-5-A Joshi Beli, *V. champiane, V. candicans, V. solonis, V. rotundifolia.*

11.5.3.5 Powdery mildew : Nistru, Bangalore Blue, Bangalore Purple, Catawba, Concord, Isabella, Champion, James, Champanel, *V. lanata, V. pariflora, V. aestivalis, V. arizonica, V. berlandieri, V. californica, V. candicans, V. lincecumi, V. longi, V. palmate, V. riparia, V. rupestris, V. tiliefolia,* Muscat Hamburg, Rangspay.

11.5.3.6 Downy mildew : *V. amurensis,* Resistant Muscat, Krivlyanskii, Muskat, Onitskanskii, Survchunskii, Bangalore Blue, Catawba, Khalili, Concord, Champion, James, *V. aestivalis, V. arizonica, V. berlandiri, V. californica, V. candicans, V. lincecumi, V. longi, V. palmate, V. riparia, V. rupestris, V. tiliefolia.*

11.5.3.7 Anthracnose : *V. parviflora,* Temple, Bangalore Blue, Muscat Hamburg, Gulabi, Phakadi, Bangalore Purple, Catawba, Concord, Champion, James, *V. lanata, V. aestivalis, V. arizonica, V. longi, V. palmate, V. riparia, V. rupestris, V. tiliefolia.*

11.5.3.8 Crown gall : V54-62, V 64-127, V 65-1, V 63-30, V 52-28, Chardeny, Gordan, Krey, Pinct, Rhine, Riesling.

11.5.3.9 *Cercospora* leaf spot: Champion, Champach,Australia No. 2.

11.5.3.10 Pierce's disease: *V. smalliana, Norris,* Lake Emerald, *V. simpsoni,* Blue Lake, *V. shuttleworthy* Pixiola, Temple.

11.5.4 Genetic resources resistance/tolerance for abiotic stresses

11.5.4.1 Cold hardiness: *V. amurensis,* Regale, Sterling *V. labrusca.*

11.5.4.2 Frost: Liana-12, *V. amurensis.*

11.5.4.3 Hot: *Vitis incecumii, V. aestivalis, V. rotundifolia.*

11.5.4.4 Drought: Dog Ridge, Ramsey, 101-14 Mgt, *V. arizonica, V. monticola.*

11.5.4.5 Soil salinity: Digraset, St. George, Salt Creek, Dog Ridge Solonis-1616, 1616-C, 216-C, 1103 P.

11.5.4.6 Chloride toxicity: Dog Ridge.

11.5.4.7 Calcarious soil: 420 A, 161-49, Hybrid 99, 110, Hybrid 150-15.

11.5.4.8 Wet soil: 101-14 Mgt, 3306C, Ramsey, Dog Ridge.

11.5.5 Genetic resources for yield and quality traits

11.5.5.1 High yield: Rubi Red, Beauty Seedless, E-26-8, E-30-14, E-29-5. Dilkush, Angur Kalan, Anab-e-shahi, Perlette.

11.5.5.2 Earliness: *V. parviflora, V. amurensis, Pearl-of-Casaba,* Beauty Seedless, Flame Seedless Perlette.

11.5.5.3 Basal fruitful bud: Perlette, Beauty Seedless, Muscat Hamburg.

11.5.5.4 High TSS: Beauty Seedless, Thompson Seedless, E-29-7, Golden Queen, Athens, Bangalore Blue, Concord.

11.5.5.5 Seedlessness: Beauty Seedless, Perlette, Thompson, Seedless, Arkavati, Arka Shyam, Arka, Neellmani.

11.5.5.6. Juice: Concord, Pearl-of-Casaba, Pusa Urvashi.

11.5.5.7 Wine: Beauty Seedless, Rubired, Black Champa.

11.5.5.8 Raisin: Thompson Seedless, Pusa Seedless, Sultana.

11.6 Ber

11.6.1 Introduction

Ber (*Ziziphus mauritiana*) is an ideal fruit plant for marginal and dry tract where other fruits can not be grown either due to unavailability of water or adverse soil conditions. The fruits are quite nutritious being a rich source of vitamin C, A and B complex.

11.6.2 Species

Species are *Z. mauritiana, Z. jujube, Z. lotus, Z. nummularia, Z.oenoplia, Z. rugosa, Z. sativa, Z. xylopyra.*

11.6.3 Genetic resources resistance/tolerance for biotic stresses

11.6.3.1 Fruit fly: Katha, Tikidi, Dodhia, Mehrun, Chinese, Sanaur-1, llaichi, Mirchia- ZG3, Umran, Thornless, Narikela, Rohtaki Gola.

11.6.3.2 Powdery mildew : Chhuhara, Safeda, Selected, Nazuk, Glory, Sanaur-2, ZG2, ZG3, ZG4, Safed Rohtak, Sanaur-5, Khatha Phal, Darakhi-1, Darakhi-2, Guli Villati.

11.6.3.3 Leaf spot: Safed Rohtak, Jhajjar Selection, ZG3, Sanaur-1.

11.6.4 Genetic resources resistance/tolerance for abiotic stresses

11.6.4.1 Cold: Chinese ber (*Z. jujube)*

11.6.4.2 Drought: *Z. jujube,* Gola, Sindhura Narnaul.

11.6.4.3 Salinity: *Z. jujube,* Banarsi Karaka

11.6.5 Genetic resources for yield and quality and other traits

11.6.5.1 High yield: Umran, Banarsi Karaka, Ponda.

11.6.5.2 Dehydration: Raja, Babu, Jeeran, Sukhawni

11.6.5.3 Long transport: Umran, Katha, Maharwali, Banarasi Karaka

11.6.5.4 Leaf as a fodder: *Z. nummularia, Z. mauritiana*

11.6.5.5 Tanning industry: *Z. mauritiana, Z. nummularia, Z. oenoplia, Z. xylopyrus*

11.6.5.6 Dwarfing rootstock: *Z. nummularia*

11.7 Litchi

11.7.1 Introduction

Litchi (*Litchi chinensis)* is the most important subtropical evergreen tree. Its fruits are famous for their excellent quality, characteristic pleasant flavour and attractive red colour.

11.7.2 Species

Litchi has three subspecies namely, *Chinensis, philippinensis* and *javenensis.* Rose scented, Bombai, Calcuttia, Late Seedless and Dehradun are imported commercial cultivar in India.

11.7.3 Genetic resources resistance/tolerance for abiotic stress

11.7.3.1 Hot and desiccating wind: Late Seedless, Calcuttia, Swarna Roopa.

11.7.4 Genetic resources for quality traits

11.7.4.1 Good quality: Muzaffarpur, Calcuttia, Late Seedless, Gulabi.

11.7.4.2 Red skin colour: Early Seedless, Muzaffarpur, Bambay, Gulabi, Dehradun.

11.7.4.3 Soft with sweet juice: Early Seedless, Rose Scented, Bombai, Calcuttia.

11.7.4.4 Pleasant flavour: Early Seedless, Bombay, Calcuttia, Gulabi.

11.7.4.5 Seedlessness: Late Seedless, Early Seedless, Small Seed Size, Nuo Mi Ci, Gui, Wei, Zhu He Li.

11.7.4.6 Early Maturity: Zhuang Yuan Hang, Yaun Zhi, , Shahi, Souey Tung, Rose Scented.

11.7.4.7 Medium maturity: Do Zoo, Fei, Zi Xiao, Hei Ye.

11.7.4.8 Late maturity: Late Seedless, Dehradun, Gulabi, Chi Ye, Huai Zhi, China.

11.8 Papaya

11.8.1 Introduction

Papaya is widely cultivated in the tropics and subtropics. It has a high nutritive and medicinal value.

11.8.2 Species

The papaya species are *Carica pentandra, C. goudotiana, C. gracilis, C. microcarpa* and *Carica erythrocarspa.*

11.8.3 Genetic resources resistance/tolerance for biotic stresses

11.8.3.1 Root rot: Wiaminalo-23, Line-8 of Solo, Wiaminalo-24, Ranchi.

11.8.3.2 Viral diseases: *C. cauliflora, C. monoica,* Pusa Majesty, *C. microcarpa, C. stipulata, C. pubescens, C. quadrifolia.*

11.8.3.3 Nematode: Pusa Majesty.

11.8.4 Genetic resources resistance/tolerance for abiotic stresses

11.8.4.1 Cold and frost: *C. candamarcensis, C. quercifolia, C. goudotiana, C. pentagona* Fair child, Punjab Sweet.

11.8.4.2 High rainfall and water logging: Pant papaya-2, *C. monoca.*

11.8.4.3 Strong wind: Pusa Giant.

11.8.5 Genetic resources for yield and quality traits

11.8.5.1 High yield: Pusa Delicious, Pusa Majesty, Pusa Dwarf, Washington.

11.8.5.2 Good keeping quality: Pusa Majesty, Co-1, Co-2.

11.8.5.3 Gynodioecious: Pusa Majesty, Pusa Delicious, Hybrid-39 and Hybrid-54 from IIHR, Coorg Honey Dew, Co-7.

11.8.5.4 Dwarfness: Pusa Nanha, Pusa Dwarf.

11.8.5.5 High total soluble solids: Solo, Honey Dew, Pusa Delicious, CO3, CO4, H-39 & H-54 (IIHR, Bangalore), Pink Flesh Sweet, Washington.

11.8.5.6 Bigger size fruits: Pusa Giant, Pusa Dwarf, Washington.

11.8.5.7 Red fleshed cultivars: Red Fleshed, Pink Flesh Sweet, Surya, Barwani Red.

11.8.5.8 High papain contents: Pusa Majesty, Homstead, CO2, CO5 and CO6.

11.9 Pineapple

11.9.1 Introduction

The pineapple (*Ananas comosus* L) is one of the commercially important fruits of India. It is a good source of vitamin A, B and C.

11.9.2 Species

The *Ananas* has five species, viz., *A. bracteatus, A. fritzmuelleri, A. comosus, A. erectifolius, A. ananassoides.* The majour groups of cultivated pineapples are Cayenne, Queen, Spanish, Abacaxi and Maipure.

11.9.3 Genetic resources resistance/tolerance for biotic stresses

11.9.3.1 Mealy bug wilt: Spanish group (Cultivars-PR-1-67, Red Spanish, Singapore, Green Selangor, Castilla, Cabezona, PR1-56, Nangka, Gandol, Masmerah, Betek).

11.9.3.2 Gummosis: Smooth Cayenne.

11.9.3.3 Root rot: Parnambuco, Red Spanish.

11.9.4 Genetic resources for quality traits

11.9.4.1 Big fruit size: Giant Kew (Cayenne group).

11.9.4.2 Canning: Cayenne group (cultivars-Smooth Cayenne, Cayenne Lisse, Hilo Cayenne, Boron, Rothschild, Smooth, Gautemalan, Typhone, St Michael, Emeralda, Kew, and Giant Kew, Champaka).

11.9.4.3 Export: Monte Lirio, Smooth Cayenne, Cayenne Lisse, Hilo Cayenne, Boron, Rothschild, Smooth, Gautemalan, Typhone, St Michael, Emeralada, Kew, and Giant Kew, Champaka.

11.9.4.4 Good flavour: Parnambuco.

11.9.4.5 Early fruiting: Parnambuco.

11.9.4.6 Early ripening: Queen.

11.10 Sapota

11.10.1 Introduction

Sapota or sapodilla popularly known as *chiku* is an important fruit. It is a rich source of digestible sugar (12-18%).

11.10.2 Species

The sapota (*Manilkhara achras)* is a member of the family *Sapotaceae.* The closely related species are *M.sieben* (Trinidad) *M. bidentata (*South America) and *M. spectabilis* (Costa Rica).

11.10.3 Genetic resources resistance/tolerance for biotic stress

11.10.3.1 Leaf spot: CO_1, Cricket Ball, Kalipatti, CO_2.

11.10.4 Genetic resources for quality traits

11.10.4.1 High TSS: CO, Kirtabarthi, Thagarampudi.

11.10.4.2 Good quality: Dhola Diwani, Kalipatti, Bhuripatti, Pala, Pot Sapota, Thagarampudi.

11.10.4.3 Good transport: Kirtabharthi, Ayyangar.

11.10.4.4 Oval fruits: Pot Sapota, Oval, Thagarampudi, Kalipatti, Dhola Diwani, Pala, Krithabharthi, Vavi Valasa.

11.10.4.5 Round fruits: Jonnavalasa Round, Cricket Ball, Calcutta Round, Baramasi.

11.10.4.6 Rose scented: Ayyangar.

11.10.5 Sapota Rootstocks

11.10.5.1 Rayan/Khirni/Pale: *Manilkhara hexandra* or *Mimosops hexandra.*

11.10.5.2 Adanis apple: *Manilkhara Kauki.*

11.10.5.3 Mahua: *Madhuca latifolia.*

11.10.5.4 Mee tree: *Bassia longifolia.*

11.11 Pomegranate

11.11.1 Introduction

Pomegranate (*Punica granatum)* is a grown commercially for its delicious fruits and juicy pink seed arils which are consumed fresh. The fruit is mainly used as a table purpose, or processed into juice which is blended with other juices.

11.11.2 Species

The family *Punicaceae* consists of a single genus *Punica* of two species, *P. granatum* and *P. protopunica. P. granatum* is indigenous to Iran whereas, *P. protopunica* is a native of the Island of Socotra, Democratic Republic of Yemen.

11.11.3 Genetic resources resistance/tolerance for abiotic stress

11.11.3.1 Tolerance to fruit cracking: Kazakai, Gulshah, Bedana, PS 75, K 3, Appali.

11.11.4 Genetic resources for yield and quality traits

11.11.4.1 Good fruit quality and yield: Ganesh, G-137, G-133, G-107, G-132, Jyothi, P-23 and P-26.

11.11.4.2 Red skin colour: Kabul, Kandhari, Spanish Ruby, Jyothi, Shirin Anar, Gulsha Rose Pink, Gulsha Red, Gulsha, Bedana.

11.11.4.3 Soft seeds: Dholka, Paper Shell, Spanish Ruby, Ganesh, Jyothi, Yercaud-1, Bedana,CO_1.

11.12 Apple

11.12.1 Introduction

Apple is the most popular temperate fruit of the north-western Himalaya region and has been under cultivation since time immemorial.

11.12.2 Species

The cultivated apples belong to the species *Malus domestica* or *Malus sylvestris* or *Malus pumilla*. Out of about 30 *Malus* species, 13 species are available in the country. *M. baccata* and *M. sikkimensis* occur wild in India whereas remaining other species have been introduced from abroad. *M. floribunda* is a source of scab resistance and its strains are also useful as pollenizer. *M. hupehensis, M. prunifolia* and *M. sikkimiensis* have apomictic seeds which can provide genetically uniform rootstocks.

11.12.3 Genetic resources resistance/tolerance for biotic stresses

11.12.3.1 Scab: Prima, Priscilla, Generos, Mac Free, Florina, Gavin, Liberty, Moira, Novamac, Jona free, Trent, Brite Gold, Red Free, Mac Shay, Murray, Pionier, Freedom, Rouvilla, Richelieu, Redsleeves, Romus 1, Romus 2, Sir Prize, Liberty, Redfree, Florina, Antonoyka, Querina, Priam, Renet Kichunova, Voinea, Prinicia, Dayton, William's Pride.

11.12.3.2 Powdery mildew: *M. robusta, M. zumi,* Koster, Khasan.

11.12.3.3 Wooly aphid: Northern Spy, Cox's Orange Pippin.

11.12.3.4 Canker: M1 and M 12.

11.12.3.5 Collar rot: MM 103, Northern Spy.

11.12.4 Genetic resources resistance/tolerance for abiotic stresses

11.12.4.1 Low chilling: Tropical Beauty, Vered, Parlins Beauty, Anna, Michal, Maayan, Rome Beauty.

11.12.4.2 High chilling: Melba, Mc Intosh, Pepin Chernenko, Diana, Antonovka, Wealthy.

11.12.4.3 Drought: M 11.

11.12 .5 Genetic resources for yield and quality traits

11.12.5.1 High yield: Greensleeves, Golden Delicious, Jester, Jupiter, Malinovka, Versified, Tropical Beauty.

11.12.5.2 Early season: Yellow Transparent, Early Shanburry, Fanny, Red June, Green Sweet, Benoni, Yandikovskoe, EC 38683, EC 32221.

11.12.5.3 Mid season: Red Delicious, Starking Delicious, Mc Intosh, Cortland, Jonathan, King of Pippin, Rich-a-Red.

11.12.5.4 Late season: Golden Delicious, Rymer, Spitzenburg, Buckingham, Ambri, Versifield.

11.12.5.5 Spur type: Starkrimson, Red Spur, Red Chief, Oregon Supr, Miller's Sturdeespur, Gold Spur, Scarlet Spur.

11.12.5.6 Red colour: Red Delicious, Red Chief, Jonathan, Empire, Red Gala, McIntosh, Mutsu, Tydeman's Red.

11.12.5.7 Yellow colour: Golden Delicious, Yellow Transparent, Cortland, Eln Shemer, Beverly Hills Dorsett Golden.

11.12.5.8 Green colour: Green Sweet, Granny Smith, Rhodels Greening.

11.12.5.9 Good shelf life: Ambri, Granny Smith, Versifield, Golden Delicious

11.12.5.10 Juices: Granny Smith.

11.12.5.11 Culinary: Newton Wonder, Antonovka, Early Victoria, Golden Noble, Rome Beauty.

11.12.5.12 Dwarfing: M9, MAC9, M27.

11.12.5.13 Semi-dwarfing: M7, M2, MM106.

11.12.5.14 Vigorous: M16, M12, M25, Novole.

11.12.6 Pollenizing varieties: Tydeman's Early, Red Gold, Golden Delicious, Mc Intosh, Lord Lambourne, Winter Banana, Granny Smith, Starspur Golden, Golden Spur, Gala, Worcester.

11.13 Pear

11.13.1 Introduction

Pear is a temperate fruit which is grown in both temperate and subtropical regions of India. It is primarily grown in Himachal Pradesh, Jammu and Kashmir and Uttaranchal.

11.13.2 Species

The primary species of genus *Pyrus* have been classified into five groups viz., European group (*P. communis, P. nivalis, P. cordata, P. caucasica),* North Africa group (*P. longipes, P. gharbiana, P. memorensis),* West Asian group (*P. syriaca, P. elaeagrifolia, P. amygdaliformis, P. salicifolia, P. glabra, P. regelii),* East Asian group (*P. pyrifolia, P. kansuensis, P. ussuriensis, P. hodoensis) and* Asian group (*P. calleryana, P. betulifolia, P. fauriei, P. dimorphophylla, P. koechnei, P. pashia).*

Commercially pear cultivars are classified into three groups, namely, European pear *(P. communis),* Asian pear *(P. pyrifolia)* and the hybrids of these two species. The important cultivars of European pear are

Bartlett, Max Red Bartlett, and Flemish Beauty while Asian pear cultivars are Sui Lee, Nijisseki and Ba Li.

11.13.3 Genetic resources resistance/tolerance for biotic stresses

11.13.3.1 Fire blight: *P. pyrifolia, P. ussuriensis, P. betulaefolia,* Ba Li, Sui Li, Le Conte, Magness, Kieffer.

11.13.3.2 Leaf blight and fruit spot: Beurre Del, Clapps Favourite, Bartlett, *P. ussuriensis.*

11.13.3.3 Pear psylia: *P. fauriei, P. calleryana.*

11.13.3.4 Woolly pear aphid (rootstock): *P. pashia,* Shiara.

11.13.4 Genetic resources for yield and quality traits

11.13.4.1 High yield: Bartlett, Flemish Beauty.

11.13.4.2 Pink skin colour: Red Bartlett, Red Comice, Stark Crimson, Max Red Bartlett.

11.14 Peach

11.14.1 Introduction

Peach (*Prunus persica*) is a very popular temperate fruit in the world. It is a rice source in proteins, minerals and vitamins.

11.14.2 Species

Three wild species, namely *P. davidiana, P. mira, P. kansuensis* and *P. ferganensis* are particularly related to peach.

According to geographical distribution, the peach cultivars have been divided into three groups namely, Northern, Southern and European or Persian group. Peach cultivars can also be divided into two groups, high chilling and low chilling on the basis of their chilling requirements. Low chilling cultivars developed in Florida during last three decades have become very popular in the submountainous Himalayan region.

11.14.3 Genetic resources resistance/tolerance for biotic stresses

11.14.3.1 Tree borer: Goldray, Elberta, Dixired, Golden Jubilee.

11.14.3.2 Canker: The cultivars derived from Elberta parentage.

11.14.3.3 Mildew: *P. ferganensis.*

11.14.3.4 Leaf curl: *Amygdalus petummikowii.*

11.14.3.5 Root knot nematode: Nemaguard, Nemared.

11.14.2.6 Bacterial leaf spot: Earligrande, Florida King, Harland, Textar, Allgold.

11.14.2.7 Gummosis: Harbrite, Harken, Harmony, Redskin.

11.14.4 Genetic resources resistance/tolerance for abiotic stresses

11.14.4.1 Low chilling: Honey, Peento, Dwarf Australia, Red Ceylon.

11.14.4.2 Cold Hardiness: Texstar, Bella, Calora, Commanche, Sunhaven, Redskin.

11.14.5 Genetic resources for yield and quality traits

11.14.5.1 High yield: Florida Grande, Sun Fire, Flordasun.

11.14.5.2 High quality: Florida Grande, Sun Fire, La Feliciana, Redhaven.

11.14.5.3 Earliness: Champion, Candor, Stark Earliglo, Flordasun.

11.14.5.4 Lateness: Gold Drop, Elberta.

11.14.5.5 Low chilling: Early Amber, Floridaking, Floridaqueen, Saharanpur Prabhat, Sharbati, Tropic Sweet.

11.14.5.6 Large fruit size: Stark Early White Giant, Fiechengteo, Halberta Giant.

11.14.5.7 High shelf life: Parrot Delux.

11.14.5.8 Nectarines: Sunred, Sunrise, Nectared 4, Cherokee; Sun Grand.

11.15 Plum

11.15.1 Introduction

Plum is a delicious juicy temperate fruit, which can be grown successfully under varied climatic conditions. It is consumed fresh as well as in dried form.

11.15.2 Species

Five species of plum viz., *Prunus domestica (*European plum) in Europe, *P. insititia,* (Damson plum) in Western Asia, *P. cerasifera* (Cherry plum) in Western and Central Asia, *P.salicina* (Japanese plum) in China and *P. americana* (American plum) in North America have different centres of diversity.

Most of the cultivars of plum belong to two different species, viz., *P. domestica* (European plum) and *P. salicina* (Japanese plum). The other important species for breeding and rootstock purpose are *P. spinosa* (Black Thorisloe). *P. cerasifera* (Myrobalan plum), *P. ussuriensis* (ussurion plum) *P. nigra* (Canada plum) and *P. Americana* (American plum). Some wild species, namely P. *cerasifera, P. umbelleta, P. undulata, P. wallichii* and *P. jenkinsii* are also found in India.

11.15.3 Genetic resources resistance/tolerance for biotic stresses

11.15.3.1 Bacterial leaf spot: The cultivars derived from American species, Damson, Green Gage, Simon, Brice, Campass, Munson.

11.15.3.2 Stem cankers: Burbank, Satsuma, Kelsey.

11.15.3.3 Leaf and fruit spot: Burbank, Mariposa, Santa Rosa.

11.15.4 Genetic resources resistance/tolerance for abiotic stresses

11.15.4.1 Cold hardiness: *P. ussuriensis, P. nigra, P. americana*

11.15.4.2 Low chilling: Golden Lalri, Kala Amritsari, Titron.

11.15.4.3 Wet soil and drought: Myroblan and Marianna seedlings.

11.15.5 Genetic resources for yield and quality traits

11.15.5.1 High yield: Italian Prune, French Prune, President, Giant Prune, Settler.

11.15.5.2 Green colour: Green Gage.

11.15.5.3 Yellow colour: Golden Drop, Golden Transparent, Kanto.

11.15.5.4 Early ripening: Green Gage, Red Beauty.

Chapter 12

Commercial Flowers

12.1 Rose

12.1.1 Introduction

Rose is the most popular flower in the world. It occupies important position among the cut flowers and is used for garden decoration and oil extraction.

12.1.2 Species

The species, namely, *Rosa brunnoni*, R. *eglanteria*, *R. foetida*, *R. gigantea*, *R. involucrata*, *R. leschenaultiana*, *R. longicuspis*, *R. macrophylla*, *R. moschata*, *R. rubiginosa*, *R. sericea*, *R. walpoleana*, *R. sericea*, *R. gallica, and Rosa chinensis* are found in India.

12.1.3 Genetic resources resistance/tolerance for biotic stresses

12.1.3.1 Mildew: Carrousel, Duet, EI Capitan, Golden Showers, Golden Slippers, Invitation, Pink Favourite, Pink Parferit, Super Star, Suspense.

12.1.3.2 Black spot: Allgold, Chanelle, Eden Rose, Korona, Peace, Pink Favourite, Stella and Superstar.

Some of the wild species have been reported to possess genes for resistance to powdery mildew, black spot and rust diseases, i.e., *Rosa rugosa*, *R. bracteata*, *R. glabra*, *R. blanda*, *R. fedtschenkoana* and *R. namothamnus.*

12.1.4 Genetic resources resistance/tolerance for abiotic stress

12.1.4.1 Resistance to high temperature: *R. clinophylla*, *R. bracteata*, Montezuma Grussen Teplitz.

12.1.5 Genetic resources for quality traits

12.1.5.1 Flower Colour

12.1.5.1.1 Hybrid Tea

12.1.5.1.1.1 Large flower: Sovia Meillard, Illonia, Red Success, Baccara and La Minuette.

12.1.5.1.1.2 Small flowers: Motrea, Mercede, Red Garnetle and Carol.

12.1.5.1.1.3 White flowers: June Bride, Matterhorn, MacGredy lovry, White Delight, Nun, Tushar, Bouche, Pascali and Virgo.

12.1.5.1.1.4 Pink flowers: Confidence, Maria Callas, Duet, Pace Maker, Happy Days.

12.1.5.1.1.5 Yellow flowers: Golden Giant, Ganga, MaGredy's Sunset and Buccaneer.

12.1.5.1.1.6 Orange-Scarlet flowers: Aruna, Fragrant Gland, Queen Faboila and Superstar.

12.1.5.1.2 Floribunda

12.1.5.1.2.1 White flowers: Himangini and Leeberg

12.1.5.1.2.2 Pink flowers: Dearest, Delhi Princes, Junior Miss, Queen Elizabeth and Traumland

12.1.5.2.1.3 Yellow flowers: Allgold, Fugitive and Goldmarie

12.1.5.1.3 Polyantha

12.1.5.1.3.1 White flowers: Katharina Zeimet

12.1.5.1.3.2 Pink flowers: Chatillon Rose and Echo

12.1.5.1.3.3 Orange flowers: Paul Crampbell

12.1.5.1.3.4 Red flowers: Ideal

12.1.5.2 Flower size

12.1.5.2.1 Large flower: Sovia Meillard, Red Success, Baccara, La Minuette

12.1.5.2.2 Small flower: Motrea, Mercede, Red Garnetle, Carol

12.2 Jasmine

12.2.1 Introduction

Large number of *Jasminum* species is grown for garden decoration and landscaping in India. It is a highly prized flower for fragrance and making perfume and essential oil. The different plant parts like leaf, stem, bark, root, seeds and fruits are very useful and most important ingredients of pharmaceutical beverages.

12.2.2 Species

More than 40 species of Jasmine are found in India. The major species *are J. auriculatum, J. flexile, J. calophyllum, J. multiflorum, J. sambac, J. grandiflorum, J. pubescence, J. rigidum, J. parkeri*

12.2.3 Genetic resources resistance/tolerance for abiotic stresses

12.2.3.1 Insect pests: *Jasminum flexile, J. calophyllum*

12.2.3.2 Fungal disease: *J. angustifolium, J. ubescens, J. communis J. aborescens.*

12.2.3.3 Mite: *J. auriculatum.*

12.2.4 Genetic resources resistance/tolerance for biotic stress

12.2.4.1 Low temperature: *J. dispermum, J. bignoniaceum, J. humile.*

12.2.5 Genetic resources for yield and quality traits

12.2.5.1 Yield of flower buds: *J.grandiflorum* clone 3 (Lucknow), J.G. Clone 5 (Thimmapuram).

12.2.5.2 Fragrance: *J. grandiflorum, J.sambac, J. auriculatum.*

12.2.5.3 Yellow colour: *J. nudiflorum, J. primulinum, J. humile, J. floridum, J.fruticans, J. giraldi.*

12.2.5.4 White colour: *J. Sambac, J.arborescens, J. auriculatum, J. calophyllum, J. grandiflorum.*

12.2.5.5 Pink colour: *J.beesianum*

12.2.5.6 Star like flowers: *J. angustifolium, J. parkeri.*

12.2.5.7 Oil and concrete: *J. grandiflorium* and its clone 5, *J. flexile, J. samback* var. manmathabhan, *J. auriculatum,* Long Round budded.

12.3 Chrysanthemum

12.3.1 Introduction

Chrysanthemum is a popular flower crop of great commercial importance as well as of aesthetic value throughout the world.

12.3.2 Species

Important species are *Chrysanthemum boreale, C. carinatum, C. coronarium, C. cinerariifolium, C. coccineum, C. frutescens, C. indicum, C. japonicum, C. maximum, C. marifolium, C. ornatum, C. rubellum, C. satsumense, C. sibiricum, and C. sinense.*

12.3.3 Genetic resources resistance/tolerance for biotic stress

12.3.3.1 Fungal diseases: Phillips (highly resistant) and 4 cultivars, *viz.*, Angela, Flirt, Gumti and SHG-3 and hybrid FV-14 as resistant.

12.3.4 Genetic resources resistance/tolerance for abiotic stress

12.3.4.1 Low temperature: *Chrysanthenum cinerariifolium*

12.3.5 Genetic resources for quality traits

12.3.5.1 White flower colour: *Chrysanthenum leucanthemum.*

12.3.5.2 Yellow and white bloom: *C. coronarium, C. segetium.*

12.3.5.3 Large white flowers: Beauty, Snow Ball, Premier, Willium Turner.

12.3.5.4 Large yellow flower: Chandrama, Evening Star, Super Giant, Coronation Gold.

12.3.5.5 Large red, bronze flowers: Autumn Blaze, Autumn Delight, Distinction, Alfred Simpson, Alfred Wilson.

Chapter 13

Spices

13.1 Black Pepper

13.1.1 Introduction

Black Pepper (*Piper nigrum L.)* the 'King of spices' is a native of India and has been under cultivation since time immemorial.

13.1.2 Species

Economically important species are *P. betle* (betelvine), *P. longum* (Indian long pepper), *P. retrofactum* (Java long pepper), *P. mullesua, P. cubeba.*

13.1.3 Genetic resources resistance/tolerance for biotic stresses

13.1.3.1 Phytophthora: *P. colubrinum.*

13.1.3.2 Root knot nematode: Pournami.

13.1.4 Genetic resources resistance/tolerance for abiotic stresses

13.1.4.1 Drought: Karimunda, Kalluvally, Neelamundi, Panniyur 6.

13.1.4.2 Shade: Karimunda, Panniyur 2.

13.1.5 Genetic resources for yield and quality traits

13.1.5.1 High yield: Karimunda, Aimpiriyan, Kottanadan, Neelamundi, Karutha Primunda, Subhakara, Sreekara, Pournami.

13.1.5.2 High piperine: Panniyur 2, Panniyur 7.

13.1.5.3 Regular bearing: Panniyur 7.

13.2 Cardamom

13.2.1 Introduction

Small cardamom (*Elettaria cardamomum),* the 'Queen of Spices' is one of the main exported spices of India. It is mainly grown in the evergreen rain forest of Western Ghats of South India.

13.2.2 Species

On the basis of nature of panicles three cultivars of cardamom have been recognized, *i.e.* Malabar, Mysore and Vazhukka. The panicle of cv. Malabar is prostrate, of cv. Mysore erect and that of Vazhukka semi-erect.

13.2.3 Genetic resources resistance/tolerance for biotic stresses

13.2.3.1 Phytophthora: MCC-61, MCC-12, MCC-40.

13.2.3.2 Thrips, hairy catterpillar and white grubs: Mudigree 1.

13.2.3.3 Rhizome rot and capsule borer: ICRI4, IISR Avinash.

13.2.4 Genetic resources resistance/tolerance for abiotic stress

13.2.4.1 Drought: CL 757, P 6.

13.2.5 Genetic resources for yield trait

13.2.5.1 High yield: S.No. 71, S.No. 81, CCS-1, ICRI-2, IISR Kodagu Suvasini.

13.3 Turmeric

13.3.1 Introduction

Turmeric is one of the most ancient spices used in India. India is a world leader in turmeric production where it is mainly produced in the states of Andhra Pradesh, Orissa, Karnataka and Tamil Nadu.

13.3.2 Species

Two species viz., *Curcuma longa* and *C. aromatica* are cultivated commercially for the production of turmeric. Some species namely, *C. amada, C. angustisfolia, and C. cassia* are grown on a limited scale. The other semiwild species are *C. decipiens, C. lorgengii, C. neilgherrensis, C. petiolata, C. aurantiaea, C. zeodaria.*

13.3.3 Genetic resources resistance/tolerance for biotic stresses

13.3.3.1 Shoot borer: Mannuthy Local, G.L. Puram-2, Chayapasupu, Kuchipudi.

13.3.3.2 Leaf spots: *C. aromaticas*

13.3.3.3 Leaf blotch: *C. aromaticas,* long duration types, Armoor, R.H.10, Ranga.

13.3.3.4 Rhizome rot: Survarna, Suguna, Sudhrshana.

13.3.3.5 Nematode (*M. incognita*): Guntur-1, Guntur-9, Rajanipet, Sugandham, Uppalapadu.

13.3.4 Genetic resources resistance/tolerance for abiotic stresses

13.3.4.1 Moisture stress: Three cultures 9 A and 9 B, Ca 66.

13.3.4.2 Drought: BSR1

13.3.5 Genetic resources for yield and quality traits

13.3.5.1 High yield: PC 78 (Swarna), PC T13 (Suguna), CLS 94, CLS 2A, RH 10., Sudharshana, Kanthi, BSR2.

13.3.5.2 High curcumin, essential oil and oleoresin: PTS 10 (Roma), PTS 24 (Suroma), Phule Swarupa.

13.4 Ginger

13.4.1 Introduction

Ginger is extensively grown in India because of its various uses as spice, condiment, beverage and medicine.

13.4.2 Species

The species namely *Zingibere officinale*, Z. *zerumbet* (L.) Sm., Z. *purpureaum* Roscoe, Z. *roseum* (Roxb) Roscoe, Z. *wightianum* Thw., Z. *macrostachyum* Dalz., Z. *cernnum* Dalz., Z. *capitatum* Roxb., Z. *cylindricum* Moon, Z. *lingulatum* Roxb., Z. *spectabilis* Griff, Z. *clarkia* King ex Benth., Z. *margianatum* Roxb. Z. *intermedium* Baker, Z. *chrysanthum* Roscoe, Z. *rubens* Roxb, Z. *squarrosum* Roxb and Z. *elatum* R. Br are occurring in India. Commercially, the ginger has different groups namely, Indian ginger (Cochin and Calicut ginger types), Jamaica ginger (aromatic, pleasant pungent and pale yellowish in colour), African ginger (dark, very pungent, lack of the aroma) and Japanese ginger (weakly pungent, aromatic and like Cochin ginger).

13.4.3 Genetic resources resistance/tolerance for biotic stresses

13.4.3.1 Scale insect: Wild ginger, Anamika, Turla Local-1.

13.4.3.2 Shoot borer: Rio-de-Janeiro.

13.4.3.3 Root knot nematode: Valluvanad, Tura, Himachal Pradesh.

13.4.3.4 Soft rot: Burdwan-1, Anamika, Poona, Himachal Pradesh.

13.4.3.5 Leaf spot: Maran, Kunduli, Wild ginger.

13.4.4 Genetic resources for yield and quality taits

13.4.4.1 High yield: U.P., Rio-de-Janeiro, Thingpuri, Karakkal, Maran, Suprabha, Anamika, Jugijan.

13.4.4.2 Bold rhizome: China, Taffinginva, SG-35.

13.4.4.3 High dry recovery type: Maran, Nadia, Karakkal, Suruchi, Turia Local-2, Tura, Thodupuzhe.

13.4.4.4 Low fiber type (vegetable type): China, U.P., Nadia, Bangkok, Poona, Jamaica.

13.4.4.5 High Oleoresin type: Rio-de-Janeiro, Kuruppampadi, China, Suprabha, Ernod Chernad.

13.4.4.6 High volatile oil types: Nassapattam, Sleeva Local, Ernad, Chernad, Himachal Pradesh, Karakkal, Valluvanod.

13.4.4.7 Tall plant type: U.P., Tinladium, Rio-de-Janeiro, Himachal Pradesh, China, Sleeva Local.

13.4.4.8 High tillering types: Wynad Local, Wynad Kunnameangulam.

13.4.4.9 Early duration type: Sierra-Leone.

13.4.4.10 Slender rhizome types: Suruchi, Kunduli Local.

13.4.4.11 High straw yielder: Sierra-Leone.

13.4.4.12 High root types: Nadia.

13.5 Coriander

13.5.1 Introduction

Coriander is an important seed spice of India which is utilized both as foliage and seed. It is mainly cultivated in the states of Rajasthan and Andhra Pradesh though it is grown more or less all over the country except the hill region.

13.5.2 Species

The Coriander (*Coriandrum sativum)* belongs to the family *Umbelliferae* and is a native of Mediterranean region. Its wild relatives are not available in the country but wide genetic diversity in both morphological and seed characteristic occur.

13.5.3 Genetic resources resistance/tolerance for biotic stresses

13.5.3.1 Fusarium wilt: Gwalior No. 5365.

13.5.3.2 Powdery mildew: R Cr 20.

13.5.4 Genetic resources resistance/tolerance for abiotic stress

13.5.4.1 Rainfed conditions: UD.1, UD.20, UD.21, GAU 1, R Cr 20.

13.5.5 Genetic resources for yield and quality traits

13.5.5.1 High seed yield: UD.41, PS.360, Gwalior No. 5365.

13.5.5.2 High foliage yield: CIMPO-5-33, VI, V_2, CO_3.

Chapter 14

Plantation Crops

14.1 Arecanut

14.1.1 Introduction

Arecanut is a common masticatory of the people and is an essential requisite for many religious and social ceremonies in India. East Indies group of Islands have been identified as its centre of maximum diversity.

14.1.2 Species

Five botanical varieties of *Areca catechu* have been recognized, viz., *A. catechu* var. *communis, A. catechu* var. *silvatica, A. catechu* var. *botanensis, A. catechu* var. *longicarpa, A. catechu* var. *deliciosa.* The variety deliciosa has sweet kernels of mature fruits.

14.1.3 Genetic resources resistance/tolerance for biotic stress

14.1.3.1 Yellow leaf disease: Indonesia-II, British Solomon Island-I.

14.1.4 Genetic resources for yield and quality traits

14.1.4.1 High yield: VTL-17 (Sreemangala), VTL-II (Sumangala), VTL-12, VTL-13, VTL-3 (Mangala).

14.1.4.2 Sweet Kernel: *A. catechu var deliciosa.*

14.1.4.3 Dwarf plant: Naturally occurring in Hirehalli, Karnataka, Harichal Dwarf.

14.2 Coconut

14.2.1 Introduction

The coconut (*Cocos nucifera)* is grown in more than 80 countries of the tropics and is a very valuable economic plant. India ranks third in area and production of coconut after Philippines and Indonesia.

14.2.2 Genetic resources resistance/tolerance for biotic stress

14.2.2.1 Root wilt: Thirty five West Coast Tall high yielding palms have been identified and are being utilized in breeding programme.

14.2.3 Genetic resources resistance/tolerance for abiotic stress

14.2.3.1 Drought tolerant: Java Giant, Fiji, Andman Giant.

14.2.4 Genetic resources for yield and quality

14.2.4.1 High nut and copra yield: Chandrakalpa, Banawali Green Round, Philippines Ordinary, Philippines Laguna, Fiji Tall, Fiji Longtongwan, and S.S. Green, Kerasree, Godavariganga.

14.2.4.2 High oil content: Chandrakalpa, Lakshadweep Micro Lakshaganga, VCH-1, VCH-3.

14.3 Oil Palm

14.3.1 Introduction

Oil Palm (*Elaeis guineensis*) is one of the important edible oil producing plant. It is used for culinary as well as in industrial purposes.

14.3.2 Species

Oil palm belongs to genus *Elaeis* which has three important species namely, *E. guineensis, Elaeis oleifera* and *Elaeis odora (Barcella odora)*. Three varieties of oil palm viz. Dura, Pisifera and Tenera are recognized based on fruit form. Tenera is widely cultivated all over the world.

14.3.3 Genetic resources for yield and quality traits

14.3.3.1 High yields and oil out-turn: Tenera palms of Ivory Coast, Nigeria and Indonesia.

14.3.3.2 Production of tenera: D x P crosses from Nigeria.

14.3.3.3 Dwarfness and oil quality: *E. Oleifera* from Central and South America.

14.4 Cashew

14.4.1 Introduction

Cashew is the most important edible nut of the world trade and India ranks first both in its area and production. Cashew plant is useful for soil conservation and is planted for development of wastelands.

14.4.2 Species

Anacardium genus of cashew has about 20 species which are endemic to Central and South America. Four species, *viz.*, *A. occidentale*, *A. othonianum*, *A. microcarpum* and *A. pumilum* are available in the country but only *A. occidentale* is cultivated for commercial purpose.

14.4.3 Genetic resources resistance/tolerance for biotic stress

14.4.3.1 Tea mosquito: G11/6 and VTH 153/1

14.4.4 Genetic resources for yield and quality traits

14.4.4.1 High yield: A large number of varieties have been released from different stations. The important high yielding with bold nuts are Selection-1, Selection-2, Ullal-1, NDR-2-1, Vengurla-3, Vengurla-4.

14.4.4.2 Medium to bold nuts: Madakkathara, Vengurla-3, accessions 13/2, 30/4, 50/2, 105/2, 107/3, 118/4.

14.4.4.3 High shelling percentage: Vengurla-1, Vengurla-2, Bhubaneswar-1, accessions 27/4, 69/1, 100/4, 10/2, 108/4, 135/4.

14.4.4.4 Early maturity: Accessions 27/2, 50/4, 130/3, 136/1, 145/4.

14.4.4.5 Cluster bearing: Vengurla-4, accessions, 30/2, 50/2, 57/1, 141/3, 144/4.

14.4.4.6 Compact canopy: Accession 12/311, 19/211, 20/311, 30/2, 30/3, 141/322.

14.5 Cocoa

14.5.1 Introduction

Cocoa (*Theobroma Cacao*) is a very important commercial plantation crop which is grown mainly in Kerala, Karnataka and Tamil Nadu.

14.5.2 Species

Three varietal groups have been recognized on the basis of pod and bean characters, *viz.*, Criollo, Forastero and Trinitario. Criollo cocoa has whitish or light brown fresh bean colour with mild and elicate chocolate flavours of finished product. Forastero beans are purple to deep purple producting chocolate of strong flavour. Trinitarios have intermediate characters, as these are the hybrids of Criollos and Forasteros. Forasteros are more predominant in India.

14.5.3 Gentic resources resistance/tolerance for abiotic stress

14.5.3.1 Drought: NC 23, NC 29, NC 31, NC 39, and NC 42.

14.5.4 Genetic resources for yield and quality traits

14.5.4.1 High yield: Hybrids 1-14 x IV-20, 1-14 x NC 42/94 and 111-105 x IV-20.

14.5.4.2 Weight of pod and dry beans: Amelonado x Na 33, P 12 x P2, Pa 7 x Na 32, Jerangau Red Axil.

Chapter 15

Medicinal and Aromatic Crops

15.1 Periwinkle

15.1.1 Introduction

Periwinkle is a tropical, perennial herb which grows up to 90 cm high. It has medicinal properties and used for controlling blood pressure.

15.1.2 Species

The periwinkle (*Cartharanthus roseus*) belongs to the family *Apocynaceae*. There are three variants *viz.*, rose purple flowers, white flowers and white flowers with a rose purple spot in the centre. The rose purple flower is cultivated and has higher alkaloid content.

15.1.3 Genetic resources resistance/tolerance for biotic stress

15.1.3.1 Die-back, collar and root rot: Induced autotetraploids, Nirmal (WH-40).

15.1.4 Genetic resources for quality traits

15.1.4.1 High leaf and leaf-alkaloid yielding line: EC 120837.

15.1.4.2 High root-alkaloid yielding line: IC 49581, autotetraploids.

15.2 Senna

15.2.1 Introduction

Senna (*Cassia angustifolia*) is a small perennial shrub containing sennosides in its leaves and pods. It is useful in habitual constipation all over the world.

15.2.2 Species

The genus *Cassia* contains about twenty five species but only few of them possess high cathartic properties. The species are *Cassia angustifolia* Vahl, *C. acutifolia* Del, *C. obovata*, *C. italica*.

15.2.3 Genetic resources for yield and quality traits

15.2.3.1 High yield: ALFT-2, Sona.

15.2.3.2 High sennoside content: Alexandrian senna.

15.3 Lemongrass

15.3.1 Introduction

Lemongrass (*Cympogon pendulus* L.) is cultivated for its essential oil, which is a good source of foreign exchange and is mainly grown in Assam, Maharashtra, Jammu and parts of Uttar Pradesh.

15.3.2 Species

Lemongrass is of two types, *i.e.* East Indian (*C. flexuosus* Ness ex. Steud Wats) and the West Indian lemon grass (*C. citrates* D.C, Stapf). East Indian lemongrass is considered true lemongrass (*Cymbopogon flexuosus*) because of its superior oil quality. East Indian lemongrass dividen into two types namely, red stemmed and white stemmed based on colour of culm. Red stemmed plants has chocolate to purple stems, yields oil of high aldehyde content while white stemmed type (C. flexuosus *var. albescene*) gives oil of low aldehyde percentage. West Indian lemongrass (*C.citrates)* has lower oil solubility in alcohol and lower citral content. C.*pendulus* is another indigenous species which is Jammu lemongrsss. It has high citral content but is cultivated on limited scale.

15.3.3 Genetic resources resistance/tolerance for biotic stress

15.3.3.1 Tolerance to little leaf malformation: Sugandhi, Pragati.

15.3.4 Genetic resources resistance/tolerance for abiotic stress

15.3.4.1 Rain fed condition: Sugandhi.

15.3.5 Genetic resources for yield and quality traits

15.3.5.1 High yield: Praman, Pragati, RRL 16, OD-408, RRL-39, Odakkali, Kaveri and Krishna.

15.3.5.2 High oil content: Pragati, Praman.

15.3.5.3 High tillering: Praman.

15.4 Palmarosa

15.4.1 Introduction

Oil of palmarosa is one of the most important essential oils. Palmarosa oil grass or Rusa grass contains high geraniol content.

15.4.2 Species

The botanical name of palmarosa oil grass is *Cymbopogon martini* var. *motia).* It is a native of subtropical India occurring mainly in the forests of Madhya Pradesh, Maharashtra and Andhra Pradesh and to a lesser extent in Karnataka, Tamil Nadu and parts of Uttar Pradesh.

Another form called ginger grass (*Cymbopogon martini* var. *sofia*) *is* also occurs widely in India and its yields oil of lower geraniol content. The other important species are *C. khasianus,* C. *nardus* var. *confertiflorus* and *C. jwarancusa.* The oil of Jamrosa is now a popular source for commercial production of gerniol in the country. Jamrosa is a selection from an interspecific cross between *C. nardus* var. *confertiflorus* and *C. jawarancusa.*

15.4.3 Genetic resources for yield and quality traits

15.4.3.1 High yield: RRL14, Trishna, Jamrosa, IW 31245

15.4.3.2 High geraniol content: Trishna, IW-31243, IW 31245.

15.5 Geranium

15.5.1 Introduction

Geranium is one of the important aromatic plants which yielding an essential oil. The main constituent of the oil are geranial and citronellol. The plant is also known as rose geranium.

15.5.2 Species

Pelargonium gravectens is the parent species for all the commercial cultivars. Some other geranium species cultivated in India are. *P.capitatum, P. graveolens, P. inguinans, P. peltatum, P. odoratossomum, P. lateripes, P. quercifollum P. radula,* and *P. zonale.* The allied species, *P.tomentosum, P. vitifoliuml* and *P.citeonellium* are also reported promising for oil composition.

15.5.3 Genetic resources resistance/tolerance for biotic stress

15.5.3.1 Resistance to wilt: Egyptian strain, Algerian strain.

15.5.4 Genetic resources resistance/tolerance for abiotic stress

15.5.4.1 Wet conditions: Annamalal.

15.5.5 Genetic resources for yield and quality traits

15.5.5.1 High yield: Algerian.

15.5.5.2 Oil content: Algerian, Reunion, PG-7 and PG-20.

15.5.5.3 Viable seed setting types: PG-7, Alg-4 N.

References

Dereuddre, J. (1992). Cryopreservation *of in vitro* cultures of plant cell. Cryopreservation of *in vitro* cultures of plant cells and organs by vitrification and dehydration. In Reproductive Biology and Plant Breeding, Dattée, Y., Dumas, C. & A. Gallais eds., Springer Verlag, Berlin, p. 291-300

Engelmann, F., (1997). *In vitro* conservation methods. In Biotechnology and Plant Genetic Resources. Conservation and Use, Ford-Lloyd, B.V., Newburry, H.J. & J.A. Callow, eds., CABI, Wallingford, UK, p. 119-162

Engels, J.M.M. 1990. Conservation strategies: A holistic approach. Paper presented at the *International Okra Genetic Resources Workshop at NBPGR*, New Delhi.

Engelmann, F. (1991). *In vitro* conservation of tropical plant germplasm - a review. *Euphytica* 57: 227-243,

Frankel, O.H. and Hawkes, J.G. (1975). Crop genetic resources: Today and Tomorrow. Cambridge University Press, Cambridge.

Frankel, O.H. 1989. Point of view perspective on genetic resources. CIMMYT Annual Report.

Frankel, O.H. and E. Soulé. 1981. Conservation and evolution. Cambridge University Press, Cambridge, U.K. 327 p.

Roberts, E.H. (1973). Predicting the viability of seeds. *Seed Sci. Technol.* 1:499-514.

Sakai, A., (1993). Cryogenic strategies for survival of plant cultured cells and meristems cooled to -196^{0}C. In Cryopreservation of Plant Genetic Resources: Technical Assistance Activities for Genetic Resources Projects, JICA, p. 5-30

Vavilov, N.I. (1926). Studies on the origins of cultivated plants. Bulletin of Applied Botany and Plant Breeding 16:1-245

Vavilov, N.I. (1935).The origin, variation, immunity and breeding of cultivated plants. In: Selected Writings of N.I. Vavilov. Transl. by K. Stass Chaster, Chronica Bot. 12 (1/6). 364 pp.

Zeven, A.C. and Zhukovsky (1975). Dictionary of Cultivated Plants and their centres of Diversity. Centre of Agricultural Publishing and Documentation, Wageningen, Netherlands, 219p.

References

Index